Cryptographic City

Cryptographic City

Decoding the Smart Metropolis

Richard Coyne

The MIT Press
Cambridge, Massachusetts
London, England

The MIT Press would like to thank the anonymous peer reviewers who provided comments on drafts of this book. The generous work of academic experts is essential for establishing the authority and quality of our publications. We acknowledge with gratitude the contributions of these otherwise uncredited readers.

This book was set in ITC Stone Serif Std and ITC Stone Sans Std by New Best-set Typesetters Ltd. Printed and bound in the United States of America.

Library of Congress Cataloging-in-Publication Data

Names: Coyne, Richard, author.
Title: Cryptographic city : decoding the smart metropolis / Richard Coyne.
Description: Cambridge, Massachusetts ; London, England : The MIT Press, [2023] |
    Includes bibliographical references and index.
Identifiers: LCCN 2022021507 (print) | LCCN 2022021508 (ebook) |
    ISBN 9780262545679 (paperback) | ISBN 9780262374811 (pdf) |
    ISBN 9780262374828 (epub)
Subjects: LCSH: Smart cities. | Internet of things. | Urban development—Data
    processing. | Public administration—Security measures. | Data encryption
    (Computer science)
Classification: LCC TD159.4 .C69 2023 (print) | LCC TD159.4 (ebook) |
    DDC 004.67/8—dc23/eng/20221011
LC record available at https://lccn.loc.gov/2022021507
LC ebook record available at https://lccn.loc.gov/2022021508

10 9 8 7 6 5 4 3 2 1

To NHS workers

Contents

Acknowledgments

I am grateful for the contributions of several PhD candidates who have helped me develop the themes presented in this book: Georgios Berdos, Eleftheria Dimitriadi, Negar Ebrahimi, Chong (Jason) Fu, Victoria Jones, Dorothea Kalogianni, Roxana Bakhshayesh Karam, Asad Ullah Khan, Gregg Sabacan Lloren, Adelaide McGowan, Graham Shawcross, Najwa Diyana Binti Sulaiman, Weijing Wang, Guodong Yang, and Chuxuan Zhang in the Edinburgh School of Architecture and Landscape Architecture at the University of Edinburgh.

Students on the MSc in Design and Digital Media and MSc by Research in Digital Media and Culture contributed extremely helpful insights as they delivered blog posts and responses to questions about social media, digital security, artificial intelligence, ethics, and the smart city.

The work also benefited from discussions with colleagues John Lee, Miguel Paredes, Andrew Connor, David House, Shameel Mohammed, Denitsa Petrova, Nick Prior, Jules Rawlinson, Agnese Sile, Harry Smith, Chris Speed, Bridgette Wessels, and especially Tolu Onabolu who first introduced me to the implications of blockchain technology, and Aggelos Kiayias, Pedro Jacobbety, and fellow researchers in the Interdisciplinary Blockchain Group in the School of Informatics at the University of Edinburgh. This work has also benefited from my association with the Blockchain Technology Laboratory and the IO Research Hub.

Some of the research leading to this book was also supported by the Centre for Future Infrastructure in the Edinburgh Futures Institute at the University of Edinburgh.

As ever, I am grateful to Adrian Snodgrass for introducing me to the rich body of ideas and sources through which to frame questions posed by digital technologies, and pioneer in digital architecture John Lansdown who encouraged me to keep writing.

Introduction: The Urban Cryptographer

In this book I want to bring cryptography into mainstream thinking about cities. Cryptography supports data security and privacy online and serves as a remedy against cybercrime and information leakage. Cryptography also supports cryptocurrencies, the blockchain, non-fungible tokens (NFTs), smart contracts, and other digital innovations that permeate the so-called "smart city." Cryptographic methods and technologies at times appear exotic and external to the concerns of those of us interested in the history, design, and shaping of cities, but in what follows I will demonstrate that cities are already invested in cryptographic ideas and practices. At the very least, cities and cryptography have concepts and procedures in common.

Without cryptography, communications among people and digital devices would be exposed for anyone to see, hack, and misdirect. Facilities for securing transactions are critical elements in city infrastructures. Cryptography applies procedures and algorithms to transform texts, pictures, audio, files, and information flows so they can be read only by a targeted recipient, a designated receiver. Cryptography is not new to the city. As long as cities have existed, communications would circulate, often in full sight, but with their messages hidden. I wish to claim cryptography as a major component in the perennial life of any city.

Formal and informal secret signals, sometimes described simply as "codes," flourish in urban contexts. Think of a knock at the door, the subtle inflection of which is known only to the conspirators, partygoers, or lovers on either side. I think also of how people exchange cryptic text messages, or the coded calls to children and pets that dinner is ready. The urban lifeworld is infused with abstruse acoustic signals that strengthen invisible connections and define spaces.[1] In 2015 an anonymous post on the social news forum Reddit invited readers to provide examples of codes that only "people in the know" would recognize.[2] "Code black" announced over the public address system tells security staff there may be a bomb in the building. "Inspector Sands to the control room, please" is a call to action for railway station staff in an emergency.[3] Coded signals are also visual. When people used to go door to door asking for handouts, we would hear of so-called "hoboglyphs," chalk marks on fenceposts of suburban houses placed there to inform fellow travelers whether or not the occupant is generous. I once saw someone tipping little piles of white flour on the curbside along my street. Later I discovered that the deposits were waymarkers for an orienteering event, signals meaningful to the runners (Hash House Harriers) who later tramped down the street.[4]

Sometimes secret messages are signs directed to a particular recipient or group. There's a specific audience, and the message is of the moment: "Inspector Sands to the control room." In other cases, the target audience is specific though the timing is not. For example, utilities, inspection points, and street furniture are overlaid with texts, numbers, barcodes, and QR codes that link parts of an infrastructure to inventories and are essential signs for inspectors and technicians who happen across them in the course of their work. Those signs are undecipherable to the rest of us. We don't even notice them. Cities are also littered with remnants of signs meant for the members of a team, but not for us. Some are of historical interest. There's a "v" shape chiseled into one of the stone steps of my tenement block. Next to it is carved the number 14, presumably a remnant of an instruction during the building's construction in the 1830s.

Cities are inscribed with formal markings such as crests, insignias, Latin inscriptions, and symbols, some of which require investigation or instruction to interpret. Their involvement in the realms of cryptography is somewhat accidental; their credentials reside simply in being mysterious to those of us in a different time, place, and culture than their originators.

Not all codes fit within widely sanctioned social norms and well-ordered urban governance. Some obscure messages are delivered as protest or in-jokes concealed within street art and graffiti. Many such purposefully transgressive signals mark territory, and define regions, neighborhoods, and activity zones for particular local communities, groups, and subcultures.

Consumer culture has its own codes and rationales. In advertising, the coding extends to subliminal messaging and other covert signals amplifying the imperative to conform and consume. Ubiquitous fixed digital-display screens and their handheld surrogates, along with electronic tags and geo-location apps, incorporate many of these cryptographic signaling functions, not least in digital advertising.

Graffiti tags, obscure markings left by a stonemason, and a service code on a fire hydrant are unlikely to serve as vehicles for the delivery of covert messages of the moment: that an invasion is imminent, your Uber has arrived, let's meet for coffee at 15:00. Code experts usually apply the term *cryptography* to deliberately composed covert messaging. In these cases the cryptographic process follows particular methods and procedures and is more instrumental and systematic than I have described so far, as I will elaborate in the course of this book.

Some Cryptographic Terminology

I have mentioned "code" a few times. *Code* is a useful term, not least as it embraces multiple meanings: rules, laws, computer programs, hidden messages.[5] You can code and decode messages, but as we shall see, the terminology around cryptography is more precise. It invokes meanings related to *hidden writing*. I've also referred to messages that are *undecipherable*. According to Simon Singh in *The Code Book: The Science of Secrecy from Ancient Egypt to Quantum Cryptography*, "A cipher is the name given to any form of cryptographic substitution in which each letter is replaced by another letter or symbol."[6] The references to "letter" assume a message that appears to human beings in text format. Symbols as discrete shapes and lines corresponding in some way to numbers, characters, and letters of the alphabet can be similarly cryptographic. Filmgoers, readers of detective stories, and gamers probably associate *ciphers* with recreational challenges and mystery stories, such as Dan Brown's *The Lost Symbol* in which we read, "The first character on the pyramid looked like a down arrow or a chalice"—that was "the letter S." [7] The association with mystery stories is reasonable. Cryptography experts commonly apply the term *cipher* to pre-electronic or digital encryption requiring codebooks, procedures, and manually operated devices. The message recipient has at their disposal a means to translate the cipher back to plain text. The *cipher text* is the string of characters produced once the message has been encrypted; the *plain text* is the original message in ordinary language. In his seminal 1960s book *The*

Codebreakers: The Story of Secret Writing, David Kahn describes *cryptology* as "the science that embraces cryptography and cryptanalysis."[8] So it implies a wider field of study than the development of methods. My adoption of "urban cryptology" advocates for such a broad-ranging study in terms of the history, theory, culture, politics, practices, and shared legacies of city development and design.

As hidden writing, *cryptography* applies to a system for translating any message from plain text into a form that obscures the message for everyone except the designated receiver. A *cryptogram* is another term for a message written in this way. You can also refer to such a concealed message as an "encrypted message" or an "encryption." To encrypt and decrypt are to convert a plain text message into a cryptogram and back again. The names given to disciplines that develop and study methods for encrypting and decrypting messages include *cryptography* and *cryptology*.

Adopting some of the terminology of semiotics, a message is a species of *signal*.[9] Signals are messages that are yet to be interpreted by human or other sentient agents. In the digital domain, algorithms check, authenticate, certify, and validate digital signals by means that often involve encryption. Cryptography applies not only to obscuring plain text messages, but also to concealing, protecting, and securing any data, process, protocol, or method of access in such a way as to render it opaque to a majority of human, animal, and machine agents. The idea of substituting letters and symbols to encrypt and transmit a singular plain text message may appear quaint in the current age of ubiquitous, rapid, and multiplied processing of secure digital signals, but the same terminology applies, and throughout this book I will enlist both pre-digital and digital examples to make the case for the cryptographic city.

The *graphic* in *cryptographic* emphasizes writing. To think of cryptography as practices in secret writing, hiding and controlling access, helps my case in asserting the ubiquity of cryptography in the city. As I will examine in chapter 2, writing provides an appropriate metaphor for the generation of meaning in cities.

Some of the examples I have provided so far apply to the visible and external fabric of the city as if cryptographic writing is imprinted onto surfaces. In this book I want to show how cryptography runs deep within the structure of the city. A city is a multilayered repository of covert communications and overlays of texts.

The word *crypto* derives from the ancient Greek word *kruptos* meaning *hidden*. It serves my purpose in describing the cryptographic city to deploy cryptography as the leitmotif for anything that is hidden. Most things in the world are hidden from the senses and the intellect. That hiddenness may be inadvertent, incidental, and even accidental. Lacking capabilities or machines that are omniscient we rely on interpretation to understand our world, and only ever in part.[10] Hiddenness also raises the questions of who things are hidden from and by what means. In many cases the agencies of necessary obscurity include coteries of experts, administrative organizations, bureaucracies and security officials, the walls of which are breached only by elaborate procedures and protocols.[11]

Computer programs, algorithms, and databases are inevitably hidden from human inspection. Programs and data contain values, variables, and processes that are invisible even before they are processed by encryption software. In this sense all digital processing is *cryptic*, and by extension cryptographic. Cities are similarly made up of hidden processes. Cities are also cryptic in so far as they involve forms of writing or are understood through metaphors of writing. I argue in this book that cryptography provides a new and effective lens through which to inspect these aspects of the contemporary city.

Smart Enough Cities

In this book I will use the term *city* broadly to capture the concepts of a dense aggregation of people, activities, and infrastructures. To date, cities are also major sites of innovation. They concentrate developments occurring in suburban, peri-urban, and rural settings and facilitate their dissemination.

Although this book is about the *cryptographic* city, the technologies, claims, and challenges of the so-called *smart* city provide the backdrop to this study.[12] Smart city infrastructures, platforms, and processes deploy apps, sensors, actuators, algorithms, artificial intelligence, and encrypted flows of big data to ease access to urban functionality and ostensibly to keep us safe, informed, and in touch. A smart city uses data sourced from traffic movement, air temperature, sunlight, and myriad scanners and sensors to activate elements of the built environment: adjust traffic lights, open and close access points, activate shading, turn up the heating, direct flows through grids, and send messages to people, machines, networks, and systems. A smart public transportation network is one where schedules and real-time data are delivered on demand to smartphone users. Such a smart system enables travelers to transition from bus to train to light rail without having to wait for connections. The system adapts its information flows to the travelers' changing needs and circumstances. Effective public transportation participates in what urban geographers Mónica Degen and Gillian Rose describe as "flows of people constantly on the move."[13]

Efficient digital retail is another indicator of a putative smart city. Market competition provides strong incentives for retailers to provide anywhere/ anytime browsing, selecting, comparing, delivering, upgrading, maintaining, repairing, and recycling of consumer products. The smart city idea extends such convenience and efficiency to city sanitation, health care, education, and governance, as well as smart grids[14] and consumer monitoring of metered local and citywide water, gas, and power networks.

A smart city ostensibly filters, delivers, and displays information for the convenience of service users and decision makers. Urban critic Shannon Mattern describes a smart-city screen dashboard as "an assemblage of tickers, gauges, feeds, and widgets that register whatever is measurable and trackable within the smart city."[15] Citizens and visitors in the smart city draw on rich infrastructures for communication and exchange via their smartphones and other mobile devices. For Degen and Rose, the smartphone screen is "the interface between the corporeal body and the smart city."[16]

Not only phones, tablets, and desk computers but also everyday products and services connect via the Internet to produce an Internet of Things (IoT). These "things" include "Smart toasters, connected rectal thermometers and fitness collars for dogs"[17] according to a lighthearted summary in *Wired*, with the potential and challenges such connectivity implies.

Smart city infrastructures carry consequences for the form of cities. Consider digital commerce within the smart city. Digital transactions have undoubtedly influenced the form of cities. Online shopping has reconfigured the pattern of retail. Under the influence of digital commerce, new consumer behaviors have brought major physical retail outlets to a close, and spawned others, such as checkout-free grocery stores. Digital commerce has emptied some city streets of activity and introduced new urban retail and leisure practices. Changes in patterns of working from home enabled by secure digital tools adjust the footprint of offices in the city, altering the supply of desk workers who would otherwise flood into city cafes and bars during time out from work. Working from home has affected office developments and introduced new spaces such as neighborhood-based work hubs, shared working environments, and the redistribution of office support services. City visitors mediate their access to accommodations, attractions, and shops via digital platforms such as Airbnb, which in turn exert pressures that change living patterns and property markets, not to mention supply, delivery, and manufacture further upchain. Add to this digital context self-drive road vehicles and secure drone-delivery infrastructures.[18] Apart from the physical structures of cities, online activity by businesses and consumers alters the way people interact with each other and with systems and services.

Securing the Smart City

None of these smart city innovations would be possible were it not for secure financial transactions, data flows, media streaming, and communications made possible by encryption. In these and other ways, cryptography plays a crucial role in the shaping of cities.

Any highly interlinked networked system is vulnerable to exploitation. The ubiquity of the Internet amplifies the risks to the security of the smart city. After all, the Internet is an open system whose default ethos and operational mode is to "share." Early myths about Internet hackers drew on the idea that anyone could tap into the flow of data to steal and alter it. According to a review article on cybersecurity in smart cities by Armin Alibasic and colleagues, threats have moved on from solitary hackers and opportunists to "highly skilled and organized professionals who are able to deploy a variety of sophisticated techniques to launch complex and coordinated attacks."[19] The researchers list these well-known bad actions as hacking and malware, zero days, botnets, denial of service (DoS), and distributed denial of service (DDoS).[20] The online book the *Cybersecurity Bible* by Hugo Hoffman presents a sobering compendium of such risks.

Cybersecurity challenges are further inflected by large centralized data hubs, the cloud, and so-called "edge computing"[21] where heavy-duty processing occurs at the site where it is needed rather than ported over the Internet. Each method introduces its own vulnerabilities.

Wireless communication adds further challenges in what one cybersecurity firm describes as the "huge and unknown attack surface on smarter cities."[22] The security firm lists targets for such urban surface vulnerabilities as traffic control, street lighting, city management, sensor arrays, public data, mobile applications, cloud and software services, smart grids, public transportation, surveillance cameras, social media, and location-based services. They show that "hacking wireless sensors is an easy way to remotely launch cyber-attacks over a city's critical infrastructure."[23] Among the vulnerabilities in each of these systems they illustrate how hackers can feed false data to exploit wireless sensor technologies and cause disruption. They provide several graphic examples: "Attackers could even fake an earthquake, tunnel, or bridge breakage, flood, gun shooting, and so on, raising alarms and causing general panic."[24] They even posit the channeling of fake data from smell or rubbish sensors causing garbage collectors to clog the streets at critical times. Cyberattacks are technical vulnerabilities that have functional, social, economic, political, and cultural effects.[25]

Smart City Narratives

By now it will be apparent that I'm moving from *smart* to *cryptographic* as the moniker of the contemporary city. But the "smart city" idea captures enthusiasms for technological innovations, many yet to be realized. The smart city also presents as a cultural *aspiration* that propels planners, policy makers, service providers and citizens toward putatively rational "data-driven" decision-making. Urban critic Shannon Mattern describes this emphasis on "whatever is measurable and trackable within the smart city" as revealing an "instrumental logic."[26] The smart city is a way of talking about the city and its challenges that adopts a certain technological gravitas. It is a formative construct in narratives about the city.[27]

The idea of the smart city is not new, though it was a concept formerly described in terms of *intelligence*. In a 1980s book entitled *Teleports and the Intelligent City*, Andrew Lipman and colleagues argued that the intelligent city would be shaped by "abundant computer-literate workers," the spread of telecommunications, "the diffusion of the personal computer in households," and "acceptance of automated decision support systems (e.g., artificial intelligence and 'expert' systems)."[28] Admittedly, the smart city

avoids some of the overly ambitious and threatening connotations of artificial intelligence.

The smart city comes across as a catch-all for a portfolio of ambitions
and initiatives. It serves in branding campaigns, and as a way of attract-
ing businesses, developers, and a workforce to a city. It carries the helpful
ambiguity that the smartness may apply to the citizens, its workforce, governance, or its technology. It also carries aesthetic connotations of spatial
neatness, order, and style.[29]

Rival discourses and metaphors compete with smart city advocacy. The
"sustainable city" is another category of city narrative.[30] Concerns about
biodiversity, climate justice, and water poverty are addressed in the sustainable city literature, but rarely problematized or addressed through the lens
of the smart city. Smart city advocacy easily assumes that digital technology
has the capacity to address sustainability and other urban challenges if not
now, then in the future. In her arguments against the rhetoric of the smart
city, Mattern declares that she's "annoyed by its elasticity, ubiquity, and
deceptiveness—and its sullying association with real estate development,
'technosolutionism,' and neoliberalism."[31]

Encrypted security measures are a key response to the vulnerabilities of
the smart city. The practical matter of security and the productive links to
urban semiotics led me to favor the term *cryptographic city* in place of *smart
city* throughout this book—or at least to test the limits of a cryptographic
framing of the urban condition.

Crypto-skepticism

Cryptography can also present in a negative light. It is easy to think of
encryption as a necessary vice that locks cities down, rendering them imper-
vious, constrained by esoteric security protocols. In so far as cryptography
translates plain text into something less readable, it denies the richness of
language and communication.

An encrypted signal gains and delivers nothing by its circulation. It
acquires no nuance, accretes no new meanings. It is meant to be incomprehensible anyway, except to a designated recipient, and bypass those communities, cultures, and influences that lie between the originator and the
targeted recipient. Encryption reinforces an exclusive and elite relationship

between sender and receiver, subjecting others to the noise of unreadable signals. The automated translation of a plain text message into a secret code and back again involves little human interpretative capability. Translating a text message into and out of its encrypted format is procedural and algorithmic.

We might therefore think of cryptography as devoid of humanity. A message gets translated into code, which is then unpacked by an agent receiving it and according to an algorithm. There is no scope for nuance or interpretation in the process. For critics, contemporary digital encryption epitomizes the restrictive, mechanistic, dehumanized, and inauthentic commerce that is digital communication.

The word *cipher* is the Arabic and Old French word for *zero*, also defining a person of no significance. Carl Jung wrote, "If . . . I despise myself as merely a statistical cipher, my life has no meaning and is not worth living."[32] By *statistical cipher* he meant the human being turned into a number. That is a common trope characterising the targets of statistical aggregation, bureaucracy, militarism, consumerism, and confinement. In this light, cryptography as writing in cipher is yet another example of this reduction of our humanity.

Cryptography also signals social failure, demonstrating mistrust. For critic Shoshana Zuboff encryption is an admission of defeat in the face of a pervasive lack of trust in the organizations, technologies, and state instruments that manage our data. She writes about how digital platforms harvest and exploit our personal data. As data is being drained unceasingly from our interactions with one another we find that encryption is the only recourse left "when we sit around the dinner table and casually ponder how to hide from the forces that hide from us."[33] She's referring to the way social media platforms harvest our personal data and sell it to others. The only way to protect ourselves is to make our communications incomprehensible to those who would exploit it. Cryptography underscores mistrust, even as it seeks to alleviate it.

A focus on cryptography also heightens other anxieties. A practical book by cryptography and cybersecurity researcher Keith Martin bears the title *Cryptography: The Key to Digital Security, How It Works, and Why It Matters.*[34] He alerts us to the pitfalls of not taking cryptography seriously: "If you are inadvertently too trusting when opening an unsolicited link to an amusing video of a dancing sheep, then your computer could easily be inducted into a global network of machines conducting criminal activities. Your computer might end up attacking mine."[35] Such concerns present as an undercurrent of anxiety to our otherwise unfettered use of networked communications. I will leave it to security advocates to elaborate such warnings and exhort computer users to engage more with the topic and to exercise increased care.

Though I am intrigued by the subject, I recognize that not everyone is fascinated by the allure of code. Cryptography can be highly technical and mathematical and is a topic many are willing to leave to the experts. We might have to attend to cryptography when reminded to choose a more secure password, and to be cautious when disclosing information about ourselves, and when parting with money. But we don't need to know its methods to do so. As with any technical study, cryptography as a topic is subject to the vagaries of taste as much as knowledge and competence.

In a helpful book on the history and legacies of cryptography, Katherine Ellison defends cryptography against some of the charges I've outlined. She agrees: "A cipher must be precise, and it must follow rules agreed upon by correspondents."[36] But in answer to the crypto-skeptics, she argues that the history of cryptography is woven into the history of emotion, and calamity, it's the last desperate resort in addressing or expressing "loss or impending death, defeat, and struggle."[37] The art and practice of cryptography also demonstrates that "ambiguity, imprecision, and flexibility characterize human language even when it is ciphered."[38] From a historical perspective she shows that skills in the art of cryptography are human and fallible: "Cipherers and decipherers do make mistakes."[39] She maintains that the stories of these imperfections can inform cultural history. She reminds us of the rich language practices that contribute to the message that is to be encrypted and the meanings and actions that it may invoke in the recipient: "Communication in cipher also requires communication not in cipher, before the cipher, when the rules are put in place and shared, when the message is decided upon. And this communication, in turn, is passed on orally or, in the case of the cryptography manuals, in print."[40]

Something similar applies in the case of telephone and video communication. Electrical impulses and binary signals between sender and receiver pick up no nuance along the way. However, the technological system is at the service of the human agency at either end of the communication channel, which has the potential to participate in the full richness of human language and sociability.

As further illustration of the cultural values evident in cryptographic practices, writers on cryptography throughout history provide awkward and deviant examples. For Ellison, "humor and playfulness characterize the history of cryptography just as much as the drive toward perfection."[41] The study and history of encryption in any case is not always serious. I surmise from her arguments that cryptography is as old as literacy. As a semiotic practice it is hardcoded into the idea of language. In her study of early cryptography manuals, she observes that "the interpretive complexities of both rhetorical and metaphorical language and of language as always coded, and knowledge of the ways in which human languages can and cannot express experience, are central to ciphering and deciphering."[42] Another way of affirming the cultural value of encryption is to observe that ciphers are as old as secrets, which society could not function without.

Cryptography is also at home within the human propensity to identify and solve puzzles. To decrypt an encrypted message without the benefit of a manual or key constitutes a puzzle. It requires that a human codebreaker work quickly and efficiently through a series of combinations. That's a further link between the city and cryptography. Cities are sometimes like puzzles that visitors and residents must "solve" as they go about their business, such as navigating from home to the shops and back. Cities are also spatial arrangements of elements (zones, parks, buildings, neighborhoods), the design of which constitutes an open-ended puzzle for designers, architects, planners, and engaged citizens. In fact, the challenge of cryptography in the city is less a problem of coding and hiding as decoding and exposing. *Accessibility* is the watchword of contemporary, culturally informed urbanism: laying bare rather than hiding things away.

The Accidental Codebreaker

Cryptography conceals and then reveals a signal to the designated receiver. To decrypt or decode is to unhide. Cryptanalysis is the difficult process of uncovering a secret message without the entitlements of the intended recipient, and without a codebook or key. Cryptanalysis is a form of breaking and entering, and often referred to as *codebreaking*. Stories of the World War II decoders at Bletchley Park focus less on the clever methods employed by the Germans to encode messages to U-boats.[43] It is the various means by which the Allied codebreakers (cryptanalysts) cracked the code, and their methods for sustaining that subterfuge, that captures the imagination. The invention of a secure coding system involves a substantial intellectual effort. But once created, it takes more intellectual and material resources to break the code than to implement it.

Codebreaking applies the endemic human propensity to enjoy and expose secrets. Codebreaking drives the human condition, more so than hiding information. It is the main ingredient of scandal and gossip, which historian Yuval Harari argues has helped homo sapiens maintain their sense of community.[44] Gossip is a way of sharing intelligence and of bonding. By our natures we seek out, as detectives and forensic scientists.[45] The prevalence of conspiracy theories provides further evidence for the priority of forensic-style interpretation. Even when there's nothing to find, we keep looking, like conspiracy theorists convinced of extraterrestrials or stolen election ballots.[46] The search, breaking the code, is sustained within the shared imaginations of communities, however fevered. Whether by intention, habitual storytelling, or accident, we are codebreakers, a human faculty that assumes even greater importance as we think of urban environments as already "coded."

Urban Affordances

Encrypted communication systems and cryptographic methods *afford* certain opportunities and risks for the city. Throughout this book I will draw on the concept of *affordance,* so it warrants some explanation at the outset. The term *affordance* was adopted by the psychologist James J. Gibson (1904–1997): "The affordances of the environment are what it offers the animal, what it provides or furnishes, either for good or ill."[47] He introduced the concept by referring to the relationship between the nonhuman animal (water bugs and bears) and its environment, though he and others also developed the concepts in the relationship between people and designed artifacts.[48] Affordances are properties of an object that present themselves as we interact with it. The doorway in a building provides an obvious example. A surveyor might measure the doorway in terms of its dimensions. But an assessment of the affordance of the doorway depends on who might use it. For someone with full mobility the doorway affords convenient passage from one space to another in the building, for someone in a wheelchair it may afford an impediment to movement, especially if there's a step, the doorway is narrow, or the door swings toward the traveler. Gibson affirms: "As an affordance of support for a species of animal, however, they have to be measured relative to the animal. They are unique for that animal. They are not just abstract physical properties. They have unity relative to the posture and behavior of the animal being considered. So an affordance cannot be measured as we measure in physics."[49] Other terms come to mind such as *properties* and *qualities*, but properties are abstract, and qualities imply a value judgment. Affordances are situated. To describe the world in terms of affordances puts the focus on experience.

The concept of affordance is relevant in the digital domain, in particular in user experience (UX) design.[50] Designers might speak of the affordance of a dialogue box on the screen for entering a password, the buttons on a smartphone, an emoji, or indeed a whole software package, platform, or service. Concepts and framings also carry affordances. In subsequent chapters I will demonstrate the relationships between affordance, encoding, and encryption.

Urban Cryptography

The chapters that follow will examine city life through the frame of cryptography. Here are some of the affordances of framing the city through an "urban cryptography." The first and most obvious affordance is that cryptography addresses challenges of hiding and securing data and information flows in the city. People, objects, and information hide in cities. By inspecting the city through the lens of cryptography, we understand better the hidden aspects of city life and form. In this case cryptography provides the urban scholar with a pretext for talking about urban hiddenness. A second affordance is that cryptography controls access. The urban challenge of cryptography is to decode and expose rather than secrete its operations within code. Contemporary urbanism celebrates accessibility, which is to lay bare and reveal to all rather than to hide and conceal. Third, cryptography exploits and relies on the properties of profligate combinations. Cryptographic processes mirror and inform what happens in the design, management, and use of city elements and spaces. In his book *The Culture of Cities*, Lewis Mumford identified cities as places where "the goods of civilization are multiplied and manifolded."[51] Cities are also places where combinations of elements, spaces, places, buildings, and infrastructure are enumerated and tested. I discuss these and other affordances in the rest of this book.

There is a strong linguistic thread to this discussion. Cities rely on and perpetuate sign systems (semiotics). As a branch of semiotics, cryptography develops further the city's place in an economy of signs. In the course of this investigation I will also deploy some of the strategies of design methods and systems theory by recasting them as tactics in cryptography. As an example, rather than dismissing the idea of design as puzzle solving I place it in a more pragmatic light, casting the designer as urban cryptographer.[52]

I will expand on the cultural aspects of cryptography in the chapters grouped into four sections. The first makes a case for an urban cryptography and emphasizes cultures of writing. The second introduces the combinatorial aspects of making, writing, and reading the city. The third delves into algorithms and calculations. The fourth and final section progresses through some extreme ventures that position the cryptographic city in temporal, global, microscopic, and extraterrestrial contexts.

1 The Case for an Urban Cryptology

1 Written in Stone

People with an eye for architectural detail can identify messages on buildings, and in parks, pavements, and cemeteries. Some of these messages are in code as inscriptions, symbols, and geometrical shapes. As persistent relics from the past, stones seem to harbor secrets. Messages inscribed on stone present as if secrets to be deciphered. Such lithic secrecy provides a touchpoint for cryptography and place. It is the persistence of coded inscriptions in stone such as Egyptian hieroglyphics and stonemasons' marks that imprint on the human imagination the idea of coded messages from the past.

I'll focus in this chapter on institutionalized secret societies such as Freemasonry. Such societies establish historical links between cryptography and urbanism. By the end of the chapter we will have extracted what we can about secret writing that's pertinent to smart city narratives, settled the limits of the historical narrative, and prepared ourselves to advance to deeper but less obvious evidence of cryptography in the city.

Secret societies were commonly associated with *hermeticism*, a mixture of traditions that draw on the mythic figure of Hermes Mercurius Trismegistus, a philosopher of indeterminate origin and the focus of various European mystery cults.[1] One of the possible origins of the name Hermes is the Greek word ἕρμα or *herma*, a heap of stones as a way marker or a place for offerings to the gods. Hermes is also the trickster god of Greek legend.[2] The word *hermetic* has come to mean something that is hidden, exclusive, secretive, locked away and sealed. According to various readings Trismegistus is a divinity and his following may have originated in Ancient Egypt. *Trismegistus* means "thrice great," referring to this mystery figure's putative status as priest, philosopher, and king.

The hermetic tradition also draws on the art of interpreting numbers known as numerology, based on the idea that certain numbers and number combinations have significance and influence beyond their role in mathematics and calculation. According to one scholar, numerology is "the belief that things happen because numbers make them happen."[3] The writings of the ancient traditions of the Hebrew Kabbalah also draw on numerology and the occult.[4]

The French scholar Jacques Gaffarel (1601–1681) is of interest as a proponent of secret writing, and notably the alluring presentation of what he termed "the Celestial Hebrew Alphabet."[5] Such writing exploits apparent correspondences between constellations and letters of the alphabet. He seemed to present this system as an alternative or complement to the signs of the zodiac and for reading the night sky for interpretations of earthly events now and in the future. A helpful article by Arnold Lebeuf explains the authority claims behind the system:[6] "The Renaissance esoteric tradition of sky alphabet was directly influenced by the Jewish Cabbalah, mystic ideas presenting the creation of God as a text, a piece of literature, a mathematical and semantic potential of creative combinations."[7] These traditions are current in the twenty-first century—for example, in the architecture of Daniel Libeskind, whose diagrams on paper, on the ground, and on his buildings seem to draw explicitly on Kabbalist symbols.[8]

For another contemporary example deploying mystery writing consider the controversial and short-lived CCRU (Cybernetic Culture Research Unit) at the University of Warwick. The CCRU drew on philosophy, mathematics, science, and engineering to celebrate and amplify imaginative myth making—hence their neologism *hyperstition*, an amalgam of *hyper* (excessively energetic) and *superstition*. They celebrated a particular network diagram they called a "nomogram" described on a CCRU web page as a "Decimal Labyrinth" with "syzygies," "zygonovism," a "tractor zone," and "primary flows." That lexicon provides something of the flavor of contemporary numerological discourse. It is inventive, ambiguous, esoteric, cryptic, and easily dismissed by skeptics. Significantly, in support of the theme of this book, insofar as cities and built environments are designed, arranged, and modified with these hermetic, Kabbalist, and numerological traditions in mind we are entitled to attribute influence from cryptography. These traditions recur throughout history in the architecture and rituals of various secret societies.

Thinking in Stone

Freemasonry is among the secret societies that draw on mystery traditions and numerology. A book published in 1875, *The Pythagorean Triangle: The Science of Numbers* bears the imprint of Freemasonry and outlines the significance of certain numbers in the society's rites and its architecture.[9] The latter draws on the dimensions of the Temple of Solomon and other metrics that appear in the Old Testament of the Bible.

Most of us first encounter secret societies through fiction, such as in *Indiana Jones and the Last Crusade* (dir. Steven Spielberg, 1989), *His Dark Materials* (dir. Tom Hooper, 2019),[10] and Dan Brown's novels. Brown's *The Lost Symbol* takes place in the unlikely setting of Washington, DC, under the Capitol Building, drawing on the mythos surrounding its founding in a Masonic ceremony by the most famous Freemason, George Washington (1732–1799).[11]

Secrecy and mystery of course run counter to the overt messages of mainstream intellectual development in philosophy, religion, politics, science, and the arts, at least since the Enlightenment. There's no room for secret communications and hidden wisdom in an open and democratic society that seeks liberty, equality, community, and openness to the power of clarity and explanation. That explains in part the resistance within architecture, planning, and other disciplines concerned with the built environment to a wholehearted embrace of institutions that are founded on secrets. That said, secret societies grew as major elements of civil society in the Age of the Enlightenment and persist to this day, institutionalized and touching on professions responsible for the built environment, notably architects and their buildings. Societies of Freemasonry and their "lodges" were at the forefront of such secretive institutions.

The Freemasons call their meeting places and their associated communities *lodges*. That's a reference to the simple shelters in which Medieval stonemasons would work under cover from the elements on construction sites. The architectural historian James Curl provides an extensive account of the historical development of the lodge as a meeting place and its eventual transformation into a theatricalized version of the Temple of Solomon described in the Old Testament book of 1 Kings, chapter 6.[12]

Rather than recount that history here, I'll briefly consider the fortunate architectural coincidences that reside with the word *lodge*. A lodge was originally a shelter, with only simple functional articulation of parts (i.e., rooms), as in the case of a shed, pavilion, cabin, booth, or bothy. Related to *lodge*, we have *loggia*, an annex to a building that is open on some of its sides, like a porch or veranda. The lodge may be permanent like a hunting lodge, but you don't stay there permanently. Lodgings are temporary places of residence, less permanent than a house or a home, though the short-term arrangement persists in the naming of long-term accommodation. In French housing is *logement*, in Italian *alloggi*, and in Spanish *alojamiento*.

Sometimes words are associated by happenstance that contributes to their adoption. Log cabins might be lodges but there's no etymological connection between *log* and *lodge*. Beavers cut logs and build and shelter in lodges. A ship's log is derived from the old practice of recording the progress of logs strung together with rope and cast out from a ship to gauge its speed. The ship's log is a *logbook*, though at school I recall that a logbook was a book full of logarithm tables (from *logos*, ratio). That used to confuse us but reinforced the relationship between logbooks and data. To *log* information is to enter it into a logbook, ledger, or database. To *log on* or *log in* is to gain access to the database or system by entering your credentials. Then you may lodge your interest, complaints, or money, which is to deposit or place some content. So, if it seems logical to do so you can log in to lodge your payment for your lodgings. Dare I say, by this lexicographical logic the *lodge* provides a kind of early entry point to data in the city.[13]

The extensive and scholarly *Handbook of Freemasonry* edited by Henrik Bogdan and Jan Snoek[14] indulges no such word play, but is instructive on Masonic institutions. Much of architecture's theory and myth relates to the crafts of working stone, not least to measurement and to establishing ratios, and stone's affordances such as longevity and dimensional stability. According to Andrew Prescott in the *Handbook of Freemasonry*, "Freemasons were originally a specialist grade of stonemason, who specialised in the carving of freestone."[15] Freestone was fine stone that could be carved in any direction to create free forms. Freemasons were the elite of the masons and carved capitals, bosses, friezes, and gargoyles. Coteries of aristocrats, politicians, and merchants adopted the masonry guild ethos and its status and symbols. Some well-known architects were among them, including Christopher Wren (1632–1723) and John Soane (1753–1837). By most accounts, Freemasonry as an organization began in Scotland (figure 1.1). As a quick summary, David Stevenson wrote the following about the early days of

Figure 1.1
Symbols over the entrance to the Edinburgh Lodge, Mary Chapel. The building was adapted in 1893 by the Freemasons and houses minutes of meetings dating back to 1599. *Source:* Author.

freemasonry: "By the seventeenth century Scotland possessed a network of permanent institutions calling themselves lodges. Membership, at first, consisted almost entirely of stonemasons, but over time men of other occupations and social statuses were admitted, from craftsmen to noblemen. Within lodges there was brotherhood, but also a division into two ranks or degrees: entered apprentices and fellow crafts (also known as masters)."[16]

Stevenson goes on to highlight the secretive nature of such organizations: "Members had secrets, collectively known as the Mason Word, into which they were initiated by elaborate rituals. These contained references to historical traditions relating to the mason craft and lodges, and included secret recognition codes by which initiates could identify each other. Compasses and the square played a part in their symbolism."[17] The "secret recognition codes" were not electronic or digital of course but presaged the idea of encryption keys and passcodes. Credentials, codes, rituals, and symbols: these access the secrets of the lodge.

An Unsecret Society

If secrets are Freemasonry's raison d'être, then a researcher might expect the society to have some purchase in a kind of *topo-* or *arche*-cryptology, especially if it is founded on myths and rituals about a building, Solomon's Temple, and in which its *Early Masonic Catechisms* refers to God as "thou great Architect of Heaven and Earth."[18]

Being secretive puts any organization at a disadvantage. A secret society founded on the idea of "secrets which must never be written" works against developing a vibrant scholarly tradition or collective memory. Freemasonry offers texts about numbers and codes,[19] and harbors archives of writings about its history and practices, but it is left to others to advance general theories or applications of the covert, secretive, hidden, or cryptographic.

In its early days in the seventeenth century, Freemasonry incorporated the pre-Enlightenment idea of the *memory theatre*, the skills exhibited by certain orators to memorize the key points in a speech by associating them with an imaginary room layout. That's interestingly spatial and architectural, but it implies a legacy in oral tradition rather than writing—secrets passed on by word of mouth, though written texts inevitably intervene.

In spite of its fascinating history, periods of persecution, political intrigue, infighting, and influence, and a rich pantheon of adherents (Mozart, Voltaire, Goethe, Washington, Conan Doyle, Kipling) Freemasonry's main strength was that it offered the benefits of club membership in a formal setting, providing companionship at its mostly male gatherings. That's a major insight from historian John Dickie's *The Craft: How the Freemasons Made the Modern World*.[20] Participants in its anachronistic, funereal cosplay were united by the idea of secrets. After his careful historical account, Dickie offers a more casual summary of the movement in an article in *Time*: "Masonic rituals consist of secrets, wrapped in secrets, wrapped in secrets.

Once the wrapping is removed, what is revealed are moral principles of utterly disarming banality. Be a nice fellow. Learn more about the world. Remember that death puts things in perspective. The great secrets of Freemasonry are all motherhood and apple pie."[21]

Dickie shows how secrets serve as a recruiting tool. People are lured in to participate in the organization's secrets. Some are attracted to clubs that have exclusive entry criteria. Secrets also keep people together. Like families, organizations can bond by agreeing to keep their secrets, even if the secrets are just about their rituals. Dickie wrote: "Masonic secrecy is not a way of hiding anything at all. It is the wrapping, and not what it contains, that is key. Secrecy is a way of enveloping bonds of fellowship in solemnity and sacredness."[22] According to *The Early Masonic Catechisms*, a freemason may not disclose any of the masonic secrets "unless to a True and Lawful Brother."[23] That presupposes you know who a "brother" is when outside the bounds of the lodge. It is tempting to say that the secret is that there is no secret. But the main covert asset of the lodge is to know who is in it—within its fraternal embrace.

Mostly, the architecture of Freemasonry involves providing a stage setting for rituals. Scholarly books and articles on buildings that incorporate motifs from Freemasonry will frequently point out specific building elements and their referents—for example, the ubiquitous drawing compass and square found on Masonic tombs, pediments, and regalia that ostensibly reference devices for marking and measuring building elements. In his essay "Freemasonry and Architecture," James Stephens Curl elaborates: "In masonic terms, the vertical has associations with licence and the horizontal with restraint, so the square defines how the vertical and horizontal are joined in a manner that would be sound construction in a building made of stone. Thus these implements are representative of morality."[24]

The presence of two pillars on either side of a building's entrance, especially if flanked by walls in the manner of *in antis*, is a further reference to the lost Temple of Solomon. But in keeping with the tenets of a secret society the allusion could be subtle, according to Curl: "the Solomonic Temple as an idea could be alluded to by this subtle means, not overtly, but perhaps slyly, missing those who were uninformed."[25] Any building may be organized around and adorned with covert references and in-jokes. It seems that buildings under the sway of Freemasonry participate overtly in this tradition. The incorporation of symbols in art and architecture does not of necessity indicate a cryptographic ethos or mindset. We have to dig a little deeper. I will probe architecture's cryptographic substrate further as I proceed.

A Secret House

A reader of architectural history might think Freemasonry's secretiveness contributes little to an understanding of architecture and cities beyond what neoclassicism offers, and as a channel for patronage. That's apparent from the scholarly literature on Freemasonry. One prominent architect provides an exception with work that adopted principles and attitudes shared with Freemasonry. The prominent Regency architect John Soane (1753–1837) was a Freemason. In an account provided by architectural historian David Watkin, Soane "took Freemasonry very seriously."[26] Though he wasn't initiated as a member until age sixty his work adopted the attitude of Freemasonry: "He reflected its deistic philosophy in his own references to God as 'the Architect of the Universe,' and in his numerous designs for funerary monuments, tombs, sepulchral chapels, and mausoleums."[27]

Soane designed the Masonic Hall in London, which was completed in 1830, though it is no longer standing. Professionals might join clubs and societies for business contacts without incorporating the organization's rites and symbols into their professional practice. A society's moral codes, communality, and work of improvement or redemption are sufficient to exert influence on professional life. But it seems as though Soane's commitment to the ideals of Freemasonry as a secret society is evident in his architecture, even domestic architecture.

Soane purchased a terrace house at 13 Lincoln's Inn Fields in central London in 1795. He lived there with his family and extended and renovated it over a number of years so that it combined the functions of home, studio, library, and museum. The house was Soane's lifework and passion documented notably in a book by Helene Furjan, *Glorious Visions: John Soane's Spectacular Theatre*.[28] Having visited the house on two occasions, I observed it to be austere, melancholic, funereal, formal, ceremonial, and symmetrical, as might befit a homage to stonemasonry. It crams enough artifacts, adornments, and architectural features to fill a building ten times its volume. This oddly scaled building fascinates and invites curiosity.[29] Furjan's scholarly assessment of the building outlines Soane's debt to the aesthetic tastes of eighteenth- to nineteenth-century London for sentimental fiction, collections of antiquities, and theatres and spectacles, as well as to theories of the sublime, shadows, and mysterious light—motifs adopted by many architects at the time, though amplified in Soane's work.[30]

Mystery features prominently in Furjan's account of Soane's house-museum, and here there is a direct link with architectural secrets as suggested by a focus on the *crypt*, a hidden place. As we know, the word *crypt* provides the root for the word *cryptography*. Soane's crypt is open to the rest of the house, and you can look down into it from various points in the plan as an evocation of mystery. It is configured as a romanticized funerary crypt. But Furjan says the same spatial devices apply to the rooms above: dome, colonnade, picture gallery, and even the breakfast room and other domestic quarters. She notes that Soane extended quasi-Gothic, neoclassical conventions, transforming Gothic literature "into a three-dimensional, inhabitable spectacle."[31]

I like to see the building in terms of secrets, as befitted the product of a devotee of a secret society. After all, theatricality is a controlled art of revealing and concealing. Furjan relates the house-museum to "representations and constructions of landscapes,"[32] particularly in the house's association with tropes of the garden picturesque. The building employs "a series of carefully framed scenes and prospects."[33] Entrances to rooms frequently align to produce a viewing sequence (an *enfilade*). Sometimes there are glimpses encountered through "apertures" that "suddenly come into line" as you move across a room.[34]

Hidden within this account of prospect is its landscape converse of *refuge*, though that's certainly evident in the house-museum. It has nooks and hiding places. The visitor to the building frequently looks down, up, or though from a position of safety: a place from which you can see without being seen, or where you can retreat from view if you want seclusion. The requirements of a private house combined with a semi-public museum demand that the architect give attention to a play between prospect and refuge. Other aspects of the building make even more overt reference to secrecy: cupboards, cabinets, and the famous gallery of paneled walls that hinge open to reveal a series of paintings by J. M. W. Turner among others (figure 1.2).

I like to think that the house-museum by Soane is not only a place of secrets, but also a lesson in the architecture of secrets. It is fair to say Soane

Figure 1.2
Moving panels in the Picture Room, John Soane's house. Photo: Gareth Gardner.

demonstrated affinity between his secret society membership and his house as an instruction manual in the spatial art of secrets. If nothing else, the building demonstrates that architecture has more to contribute to the theory and practice of secret keeping than do secret societies. I would add the offer of crypts, basements, darkened rooms, and cupboards to the reasons secret societies gravitate toward architecture: its histories and theories, functions, types, and symbols.

Spatial Affordances

How do the tenets of cryptography interact with both architecture and the rules and practices of secret societies? Assemblages of elements, social movements, cities, societies, and systems carry *affordances*. I introduced the concept of *affordances* in the introduction as those properties, attributes, or qualities of things that are relative to the person or creature that encounters them. The steps leading up to an urban public square afford movement between levels, or they afford sitting outside if the weather is suitable. They also afford jumping and ollieing for a skateboarder, or they afford

Figure 1.3
"Architecture" in pigpen cipher. *Source:* Author.

hindrance to people who are less mobile. Though they may appear as adjectives or nouns (e.g., "accessibility," "concealment"), affordances start with actions (e.g., "to access," "to conceal"). It helps my case to identify seven affordances through which cryptography communicates possibilities for action in the city, and even within secret societies.

So, the first affordance is to *conceal*. Traditionally, cryptography deploys ciphers for concealing messages, communicated via text, symbols, or other devices, such as knots in string, abstract markings, or binary signals as in Morse code in sound, light, or other media. *The Early Masonic Catechisms* refers to cryptography; for example, a master says to his novice: "I could not avoid immediately thinking of the old Egyptians, who concealed the chief Mysteries of their Religion under Signs and Symbols, called Hieroglyphicks."[35] Egyptian hieroglyphics were a perennial stimulus for secret societies, especially prior to their decipherment with the discovery of the Rosetta Stone in 1799. The most commonly used concealing method that appears on Masonic inscriptions (in particular on tombstones) is the Masonic cipher, less elegantly named the *pigpen* substitution cipher (figure 1.3). There are many explanations on the Internet, including a website by Johan Åhlén that translates clear text to pigpen automatically.[36] The coded characters are based on configurations of lines at right angles to one another and dots. Ciphers based on straight lines and dots are easier to chisel into stone than curves. They are also easier to draw with a drawing app.

Cryptography deploys ciphers for concealing messages, communicated via text, symbols, or other devices. Concealment is also an obvious affordance of objects in space. If you put something into a cardboard box and close the lid, then it is concealed from view. Buildings conceal their contents. I've already discussed how Soane's house-museum provides a vivid illustration of the subtle architectural art of concealment. In cities concealment involves basic geometry and what George Lakoff and Mark Johnson in their seminal studies of metaphor call "the containment metaphor."[37] Cryptography hides messages inside code; buildings "hide" their contents within walls. You can "hide" in a building from the elements and enemies. In his conclusion to an interesting essay outlining the history of domestic privacy, Robin Evans remarked how architecture (at least in the twentieth century) had the capacity to contain human experience by "reducing noise-transmission, differentiating movement patterns suppressing smells, stemming vandalism, cutting down the accumulation of dirt, impeding the spread of disease, veiling embarrassment, closeting indecency, and abolishing the unnecessary."[38] These are some of the basic affordances of architecture as concealment.

The second affordance of cryptography is that it *controls access*. As well as messages, cryptography provides admission to something, or the message that it reveals offers access, as if a key to a door, cupboard, box, file, or information. Access and its denial are crucial aspects of any secret society. Whether or not such organizations deploy cryptography in earnest, the idea of access concealed via codes provides a leitmotif. Cryptography serves as a meta-symbol for the secretive organization.

Buildings specialize in controlled access, revealing and concealing. This access applies to different sensory modalities: judicious revealing of sights, sounds, smells, and what you can touch. Visitors and occupants move through doors. They also peer through arches, windows, screens, and from balconies and mezzanines. Openings, shafts, atriums, stairwells, and porous membranes provide access to sight, circulation, sounds, and airflows as exemplified in Soane's subtle use of openings and enfilades.

The third affordance is that cryptography *combines*, makes patterns, introduces combinatorial complexity. In a book chapter "Two Thousand Years of Combinatorics" the mathematician Donald Knuth defines *combinatorics* as "the generation of combinatorial patterns"[39] that crosses cultures and disciplines such as poetry, music, and religion. It is also a branch of study in modern mathematics. In his book on the history of cryptography, Simon Singh relates the term *cipher* to "scrambling": "the term *encipher* means to scramble a message using a cipher."[40] From a spatial and architectural point of view I prefer the more generally applicable concepts of *combination, combinatorial complexity*, and *assembly*.

Cryptography works with combinations of symbols as in a combination lock or other abstruse sequence. The sheer number of combinations impedes access to the message or the content—unless you have the cryptographic key. For someone without the key the message appears "scrambled." As I

will examine in chapter 4, whenever theories, myths, and folklores speak
of the arrangement of elements, then they participate in the workings of
combinatorial complexity: the plethora of possibilities for variation offered
by a multiplicity of alternative arrangements.

In the case of secret societies such combination pertains to the ordering
of components in ritual. In the *Handbook of Freemasonry*, Arturo de Hoyos
amplifies this ritualistic focus as the basis of regional differences between
rites as if they are positioned on a staircase or schedule: "The degrees of
a Rite will usually, although not always, have a numerical designation or
fixed position on a calendar or schedule."[41] De Hoyos explains this ritual
profligacy in combinatorial terms, as "Rites" that "may be arranged in a
particular sequence."[42] Such variation enables rival identities and increases
the possibility for secrets.

I've already alluded to the importance of number concepts in secret
societies. Freemasonry includes mathematicians such as Leibniz, Newton,
and Poincaré among its adherents and by appropriation the workings of
mathematics via combinatorics and number theory, which invariably deal
in sets, series, and combinations.

Buildings are made up of arrangements of elements, evident not least in
the configuration of artifacts and classical architectural elements in Soane's
house-museum. By my reading this is an aspect of architectural design that
flourished in twentieth-century modernity, though nascent in earlier traditions. According to Aristotle, with beautiful things "nothing can be added
to them or taken away."[43] Andrea Palladio (1508–1580) amplified this assertion to define beauty as a satisfactory relationship of parts to one another
and to the whole. Architecture involved the arrangement of parts, though
it afforded only limited scope for experimentation and innovation: "Beauty
will result from the form and correspondence of the whole, with respect to
the several parts, of the parts with regard to each other."[44]

Factory production, modularization, assemblies and more complicated
reflections on function and human interaction expanded architecture and
urbanism as arts of perfect arrangement to the deliberative assembly of discrete parts. Though any scholar will be quick to point out that architecture
is much more than combinations, the task of designing a building now
presents architects with configurational challenges analogous to solving
a puzzle. Sometimes these configurations shift and change in real time,
via furniture, movable partitions, lifts, moving walkways, and responsive

sensor-controlled architectures. Via the affordance of combinatorics, spatial arrangement offers challenges analogous to working through combinations in codebreaking, though rarely with a single correct or optimal "solution."

The fourth affordance of cryptography is to *follow a path, to navigate.* A codebreaker (cryptanalyst) searches through a sequence with many branching paths to explore options, one of which leads to the message. One common characterization of this search is a branching maze, a network of paths, loops, and dead-ends. Branching paths sit within the methods of mathematical procedures, derivations, proofs, and algorithms.

Secret societies invoke the labyrinth as a metaphor of process, hiddenness, and obfuscation. Snoek describes ritual perambulations in Freemason initiations following the pattern of a labyrinth: "The candidate now perambulates the lodge-room three times. Traditionally, the first and third perambulations were clock-wise while the second one went anti-clock-wise, as in the traditional form of the labyrinth (the 'Troja-castle'). The perambulations go round the 'tableau,' in English referred to as the 'tracing board', a drawing of symbols on the floor in the centre of the lodge."[45] The Troja-castle and its garden labyrinth are in Prague. W. H. Matthews's 1922 book *Mazes and Labyrinths* highlights the significance of the maze for Masons, as symbol as well as in processional rituals.[46] As I will examine in chapter 5, labyrinths afford and require expertise in navigation, further aligning cryptography with the city.

In cryptography, a codebreaker searches through a sequence with many branching paths to explore options, one of which leads to the plain text message. Programmers responsible for creating secure encryption invent algorithms that draw on similar processes of path navigation.

Any building or city can present alternative paths, or a single path as a processional journey that weaves through space.[47] Soane's house-museum offers several processional routes both horizontally and vertically, an affordance made obvious to the contemporary visitor during a busy period. The Design Methods Movement championed the arts of solving spatial problems as search processes, analogous to traversing a complicated and dynamic maze of possible actions.[48]

The fifth affordance of cryptography is the pursuit of an origin. In terms of actions, that is *to source, to return* and *to restore.* Cryptography helps preserve the concept of an original—a plain text message, an original meaning. The sense of an original applies to the immediate instrumentality of coding and decoding messages, but it also invokes appeal to a long-term legacy of messages remaining to be discovered and decoded. Remnants of past communications take on the aura of secrecy. Architectural historian Anthony Vidler refers to the stage setting of Freemason initiation rites as primitivist reenactments of a return to or reinstatement of the lost condition of Adam, the first human: "This quasi-ritualistic and allegorical stage-set demonstrates the double character of the Masonic 'return to origins,' celebrating at once a rebirth founded on primary truths and the civilized 'route of progress'."[49]

Cryptography also provides analogies for architectural history (or historicism) as a journey into the hidden past. Much of architecture's historicist theorizing is dedication to origins, a return to a first authentic primitive source—the garden of Eden, the primitive hut, Noah's ark, the Tabernacle, the Temple of Solomon, the Celestial City, Utopia.[50] Vidler confirms Freemasonry's and architecture's quests to "return to origins."[51] Soane's house-museum provides many such references to origins: the romance of the collector, the crypt, the staging of historical epochs, legacies in stone. I would add that the quest for authentic origins extends into the Primitivist strands of modern art and architecture.[52]

The sixth cryptographic affordance deals with practical outcomes, *to transact*. As with devices in written language in general, the purpose of cryptography is to preserve, but also to enable social interactions, albeit with a degree of secrecy and security. There's the communication of a message and the actions that follow, such as access, the delivery of information, and the exchange of money. A novice joining a secret society enters into a contract to uphold its rules and secrets. To adopt cryptographic messages and codes symbolizes a commitment to the society. Cryptography is a further meta-symbol of the organization's secrecy but also its sense of community.

As with writing in plain text, the purpose of cryptography is to preserve, but also to facilitate social interactions. Similarly, buildings undoubtedly protect and preserve. They are also places in which transactions and interactions take place: familial, social, economic, performative, and ritualized. Soane's house-museum was a functional home after all. Reflecting architecture's pragmatism, Evans celebrates "an architecture arising out of the deep fascination that draws people toward others; an architecture that recognizes passion, carnality, and sociality."[53] The idea that architecture and urbanism provide settings for meaningful human interactions pervades conceptions of contemporary urban design.

The seventh affordance is *to trust*. Cryptography plays on uncertainties. It may even obscure the fact that it is concealing a hidden message. You can never be sure if there is a message, or if you are confronting a hoax, a sequence of random symbols designed to give the impression that there's something there or to waste the time of a would-be codebreaker. Freemasons are required not to reveal the secrets of the society, or even the fact of its secrecy. *The Early Masonic Catechisms* asserts: "never Reveal the Secrets or Secrecy of a Mason or Masonry."[54] Don't divulge the secret, but don't let on that there is a secret. Cities are also founded on concepts of trust, not least as people engage in commerce as I will examine in chapter 8. Part of the pleasure and functioning of architecture comes from uncertainty about whether you can trust that you are in a particular space or out of it, whether you should be there or not, or you are occupying the ambiguous condition of the threshold.[55] Soane's house-museum plays on such spatial thresholds, transitions, and ambiguities.

As does architecture, secret societies offer more than I have indicated so far. People pay membership dues in formalized secret societies. Such organizations are now registered charities with buildings, financial assets, lobbying power, and influence that are likely to impact on the built environment. Secret societies presumably afford their members more than rituals and secrets. There's conviviality, education, social events, community projects, fundraising, and charitable works. Historically, political causes, resistance, and activism have also driven these societies' activity profile, though that's now proscribed in most Freemasonry lodges, which see themselves as apolitical.[56]

But there are groups, networks, individuals, and "minorities" with various degrees of organization and who at different times have found themselves operating in secret due to risks of persecution or prosecution. Unofficial enclaves of trust form across cities for mutual support, protection, common identity. and activism. Religious, pressure groups and self-help communities fall within this category. Armistead Maupin's novel *Tales of the City* has come to represent San Francisco's LGBTQ+ culture of the 1970s and 1980s and explores the diverse cultural life blood of that particular city. What was once secretive turns into conspicuous presentations of freedom and diversity in urban settings.

In summary, I've identified seven affordances of a secret society, which apply to keepings secrets and to cryptography, allowing participants in certain ways to: conceal, access, combine, navigate, restore, transact, and trust. As affordances or motifs we could translate these to concealment, access, combinatorics, path finding, the search for origins, transactions, and trust. Any architecture or urban configuration might demonstrate these affordances, Soane's house-museum providing a vivid historical reference point.

A Philosopher's House

Before concluding this chapter, I would like to test the confluence among cryptography, architecture, and language philosophy as demonstrated in the work of Ludwig Wittgenstein (1889–1951). He wrote diary entries in secret code as outlined in Dinda Gorleé's book *Wittgenstein's Secret Diaries: Semiotic Writing in Cryptography*.[57] There are several touch points with architecture and place in this account. Wittgenstein had trained as a mechanical engineer and undertook a serious foray into architecture when he designed a house for his sister.[58] Wittgenstein's earlier and only other journey into architecture came with his commissioning of a hut in Norway. It was a simple structure in the local style, but it was built according to his personal preferences and he retreated there while working in Cambridge. The site was remote, isolated and at the head of a fjord with views of the mountains.[59]

Wittgenstein's use of cryptography in personal correspondence may have been a continuation of the kind of parlor games exercised by a large wealthy family of very close home-schooled siblings. I'm inferring that judgment from Gorleé's description of Wittgenstein's early life. The cryptographic code he adopted was not especially inventive. It was a relatively simple case of a substitution cipher. Further in keeping with an interest in cryptography, Wittgenstein seemed to harbor secrets. His family was of Jewish origin though they were baptized as Catholics. He was unhappy in the army and subsequently, it seems, suffered from severe depression. Three of his brothers committed suicide. His unprepossessing Norwegian hut gives expression to the nexus of place and secrecy.[60]

Scholars have also aligned some of Wittgenstein's philosophical ideas with "mysticism," at least as defined by English intellectuals at the time.[61] As explored by many scholars, prominent architects have drawn on unconventional traditions other than the purely rational. Other exemplars include

Frank Lloyd Wright who was a Unitarian, members of the Bauhaus such as Wassily Kandinsky who were Theosophists, and even Le Corbusier the avowed atheist drew on religious symbols of sacrifice in his architecture.[62] As far as we know, Wittgenstein the temporary architect drew on none of these traditions directly, but his engagement with architecture entitles us to consider his philosophy in these unconventional lights.

Considering the cryptographic affordances discussed in the previous section, Wittgenstein's personal story speaks of concealing and revealing, and in his philosophy is activated by the enigmatic final sentence in his seminal philosophical work the *Tractatus Logico Philosophicus*: "Whereof one cannot speak, thereof one must be silent."[63] On the topic of combinations, he refers to "truth combinations" as arrangements of logical signs. His concept of language games in his later book *Philosophical Investigations*[64] also foregrounds the cryptographic affordance of combinations. The *Tractatus* follows the methods of geometrical proof, which traces a path from a foundational proposition, an origin: "The world is everything that is the case,"[65] through to more advanced propositions about language, truth, and meaning. Prior to his final statement about silence in the *Tractatus*, he describes his argument as climbing a ladder. Having reached the top the reader should throw the ladder away. The transactional affordance is revealed as "logical necessity."[66] In Wittgenstein's later writings he veers toward the active, practical, and performative aspects of language. In the *Tractatus* he deals with trust in terms of certainties and probabilities: "Certain, possible, impossible: here we have an indication of that gradation which we need in the theory of probability."[67] I'm here proposing that the life, thinking, and architecture of this particular language philosopher illustrate and parallel the affordances of a cryptographic cultural mindset grounded in revealing, accessing, combining, path following, origins seeking, transacting practically, and negotiating trust.

So far, I have emphasized the built environment in material terms, as befits an argument that begins with blocks of stone. Contrary to the limitations imposed by walls, floors, and ceilings, various devices work to supplement or confound architecture's engagement in secret keeping—not least photography, image sharing, surveillance methods, and display screens, which I consider in the chapters that follow.

In this chapter I addressed secret societies and their touch points with architecture and the built environment. I called on John Soane's famous house in London as an illustration of how architecture can demonstrate the precepts of spatial hiddenness. My purpose in the early part of this chapter was to identify the most obvious cases of architecture as a canvas onto which are written coded messages in stone, concrete, and steel. Dynamic digital surfaces and screens are more fluid and adaptable as media for overt or hidden messages that have currency in the moment, a theme addressed in chapter 3.

Chapter 2 will advance further the case that cryptography permeates and affects cities. I think of the city as an arrangement and rearrangement of components and elements: buildings, streets, neighborhoods, blocks, pipes, and services. As a process of arranging, it bears similarities to writing.

2 Write Me a City

If cryptography is hidden writing, then I need to establish that cities are already sites of plain text, in other words, ordinary language and unencrypted writing. In this chapter I will martial evidence that affirms the connection between place and writing. This provides a grounding for my case that urban and cryptographic affordances are related.

Urban Ciphers

Cryptography is an old practice embedded in the functioning of cities and nations. Generals and soldiers would pass messages up and down the chain of command in secret to avoid interception by an adversary. Writing in the 1600s, the English natural philosopher John Wilkins (1614–1672) affirmed that whether in a dungeon, a city under siege, or a hundred miles away "there are certain ways to discourse with a friend."[1] He was referring to cryptography and he recounted numerous methods for sending messages in secret, including examples of the kinds of short messages delivered by citizens and leaders under siege.[2] Epidemics and starvation were among the afflictions that beset a besieged city and required secret communications.

   I am largely drawing this historical account from the book *A Cultural History of Early Modern English Cryptography Manuals* in which Katherine Ellison recounts the four-hundred-year history of secret messaging as a necessity, a hobby, and an obsession.[3] Wilkins's book of the 1600s is a prime example of one of these early cryptography manuals. As I will demonstrate in this chapter, secret writing dates to the dawn of language. By related accounts, people's interest in cryptography was amplified in the age of the printing press in the 1400s.

According to Ellison, in Wilkins's time cryptography provided "a promising solution for loneliness and survival in a disconnected world of local and global conflict."[4] Crossword puzzles, Sudoku, and Wordle are no doubt contemporary survivors of the recreational study of ciphers and cryptography. Solving puzzles, playing video games, posting on social media, and sending messages are common diversionary pastimes that fulfill similar appetites.

There is a complete facsimile version of Wilkins's *Mercury* at Google books and elsewhere. The full title on the book cover is *Mercury: Or The Secret and Swift Messenger. Shewing, How a Man may with Privacy and Speed communicate his Thoughts to a Friend at any distance.* The book explains that the most cunning methods of passing messages were those concealing to a potential interceptor that there is even a message in play. One method involved an innocuous knotted piece of string secreted around a saddle or threaded into clothing. When unfurled and zigzagged across a wooden template guided by side notches, the otherwise random knots line up against the letters of the alphabet to reveal the secret message. The message so derived in figure 2.1 says to "beware of this bearer who is sent as a spy over you."

Not all cryptographic methods were efficient. Wilkins recounts how the sender could shave the head of a young servant, then write something on the servant's scalp. When the hair had grown back the servant would be dispatched to the recipient, who would then shave the servant's head to reveal the message. Not all messaging techniques are instantaneous. Wilkins doesn't provide an example of the kind of message that would be so conveyed.[5]

Words and Places

Though cryptography pertains to all manner of secret communications via symbols, drawings, geometries, sounds, and movements it is most at home as an art of concealing text messages. Pioneer of digital media in architecture and urbanism, William J. Mitchell wrote extensively on the cultural nexus between digital technologies and the culture of cities. In his book *Placing Words: Symbols, Space, and the City*, Mitchell makes an astute observation linking text to place.[6] Where words appear in physical space really matters, he observes. Meaning relies on context, and location is a crucial aspect of the context in which words both spoken and written assume their meanings. He provides the example of the word *fire*. Uttered in a

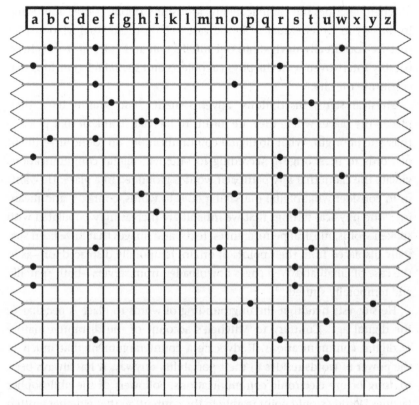

Figure 2.1
String cipher. When the string is unfurled and zigzagged across a wooden template the knots line up against the letters of the alphabet to reveal the secret message. Redrawn by the author from Wilkins's *Mercury* (1641), 46.

crowded theater it means something different than when uttered on a rifle range. Think also of the place dependence of signs, such as "No standing anytime." I have stumbled across several photographs online of someone standing on the shoulders of another person while clinging defiantly to a road sign that says "no standing on shoulder." Assuming you get the dual meaning of the word "shoulder," the joke here is achieved as if acting in defiance of the sign, as though the words belong in some other context, perhaps in a gymnasium.

The place dependence of word meaning already entitles architecture, planning, and urbanism to claim crucial roles in the way language works.

Mitchell describes the complicated relationships among people and things in the city as "intertextual": "So the vast web of intertextual relationships that we continually navigate in our intellectual and cultural lives is inextricably interwoven with the physical objects and spatial relationships that constitute the city."[7]

Physical signs, inscriptions, advertisements, and graffiti have their place—the places where they are meant to be read. Think of that place as their site of origin. What do digital communications do to the attachment of words to their origin? According to Mitchell, digital communications violate the claim that any particular word is grounded in a place. He wants us to consider what digital messaging does to words, contexts, and meaning: "the physical settings that we inhabit are increasingly populated with spoken words, musical performances, texts, and images that have been spatially displaced from their points of origin, temporally displaced, or—as in the case of email and Web pages downloaded from servers—both spatially and temporally shifted."[8]

He suggests here that the ungrounding of words to place is peculiar to the digital age. Contrary to Mitchell's account of the disruptive nature of digital communications, I advocate a counterargument that much of the power of words resides in their portability anyway. It has always been so, long before electronic communications. Followers of the philosopher Jacques Derrida would point readily to the contention that whether written or spoken, words move from one location to another, in the recounting of what someone said, memorizing and reciting prose, and as written down.[9] Rather than embark on a critique of words' dependence on place, I want to accept Mitchell's assertions as evidence that entitles us to discuss text and the city in the same breath, as it were. Texts, whether rooted in or uprooted from places, are part of the city. I also want to show how important mobility is in thinking about the shape of cities—not how people move about the city, but the city as a configuration of movable elements. Further on in this chapter I will talk about movable type in the printing process, but at this stage in the discussion I want to establish the various ways that text, writing, cryptography, and the city are related.

The quotes earlier on in this section were from Mitchell's last book on cities in a career cut short before cryptography came to the fore as a consideration in the smart city. But he observed in 2005 that under the sway of digital programming, the elements of a city can and will operate in much the same way as someone speaking to you: "programmable objects can perform speech acts, and autonomously engage you in various forms of discourse. They can query you, demand information such as passwords, refuse you access, provide you with information, accept your instructions, and issue orders to you."[10] That communicative aspect of the city also harbors the potential for deception. These digitally communicative elements of the smart city can cause harm: "They can dispense facts, fictions, and lies. And malicious computer viruses, worms, and Trojan horses can take over networked, programmable objects to do you harm."[11] By this reading, texts in the city are implicated in the challenges of security, privacy, falsity, and verification, part of the language of cybersecurity and cryptography.

Other researchers align cities to writing, though under a more general category. Urban theorist Shannon Mattern sees the city through the category of *media*. She affirms the influence of writing, voice, text, printing, and the media in which they are expressed that include steel, stone, mud, clay, wood, and paper in the formation of cities. Cities have similarly influenced writing. She positions concepts of the contemporary smart city in this broader historical and cultural context: "today's smart cities don't have a monopoly on urban intelligence. In fact, we can trace that 'smart' genome all the way back to ancient Rome, Uruk, and Çatalhöyük."[12] Mattern is referring to the centers of Roman, Mesopotamian, and Neolithic influence.

The metaphor of city as medium is powerful in the context of cryptography, Mattern states: "For millennia, our cities have been designed to foster 'broadcast'; they've been 'wired' for transmission; they've hosted architectures for the production and distribution of various forms of intelligence and served as hubs for records-management; they've rendered themselves 'readable' to humans and machines; they've even written their 'source code,' their operating instructions, on their facades and into the urban form itself. They've coded themselves both for the administrative technologies, or proto-algorithms, that oversee their operation and for the people who have built and inhabit and maintain them."[13] Her reference to codes and algorithms helps ground further the framing of the cryptographic city.

There are other connections between writing and the city. Cities often confuse visitors, as if they harbor secret messages. The texts of any city are not always clear and unambiguous. Consider the challenge of reading and interpreting the artifactual remnants of a ruined city such as those of Rome,

Figure 2.2
Monte Albán, Valley of Oaxaca, Mexico. *Source:* Dmitry Rukhlenko via Shutterstock.

Uruk, or Çatalhöyük, especially where its inscriptions and graphic symbols are undecipherable. While researching cryptography in cities, I came across an article by Robin Heywarth about Monte Albán, the ancient Mexican city (figure 2.2). The article calls Monte Albán the "encrypted city." That phrase set me thinking about the extent to which any city can be so described. The ruins of Monte Albán are marked with so-far undecipherable symbols and markings etched into the stone fabric of this ancient city. The article also claims that the lines of the city's geometry are similarly coded: "Whilst the lines alone could be dismissed as meaningless . . . the numbers of proposed alignments add weight to the idea the city is encrypted with astrological information that would be easily deciphered by the High Priests of the city."[14] Ancient cities will inevitably appear as if in code, and encrypted, especially if they suffered decline or were destroyed before European scholars had a chance to tap into local memory pools. As I discussed in chapter

1, it is easy to regard hand-carved stone remnants as speaking from the past. But like the symbols in the ancient city of Monte Albán, digital cryptography writes itself across any city. Dare I say it even *writes* the city.

Shibboleths

Not only writing, but also the way things are said links text to places. You can tell where someone is from, or not from, by asking them to say a particular word. It can also indicate where someone has been. I can tell with a degree of certainty if someone has been to Australia by the way they say "Melbourne." If they draw out the final syllable, as it's written, then they have probably not said it often enough and in the right company to be "corrected" to say "Melb'n."

A shibboleth is a similar kind of pronunciation test. The word *shibboleth*[15] comes from a story in the Biblical book of Judges 12:6. The word simply means "the ear of a grain." When prompted to say "shibboleth," someone under suspicion wouldn't be able to pronounce it with the full "sh" sound as would a native speaker of the Hebrew dialect and so would give themselves away as the adversary. There's a spatial aspect to the use of words in this way. Words sound different according to where they are spoken and by whom. Dialect belongs to place. Something similar applies to vocabularies. The ancient Greek hero Odysseus was told to carry an oar from his ship with him to the center of the island of Ithaca.[16] When people he met started referring to the oar as a *winnowing fan* (for separating wheat from chaff), then he knew he was a long way from the ocean.

Some activists now recruit the word *shibboleth* to draw attention to racial and cultural division. *Shibboleth* was the name of an art installation by Doris Salcedo at Tate Modern in London in 2007. The artist hammered a big crack in the floor of the main exhibition hall. The crack has since been filled in with cement and rendered over to create a smooth but still visible trace. Doris Maria-Reina Bravo provides a helpful post explaining the naming of the work: "Every community, culture, and nation has its shibboleth. Among the U.S. military, 'lollapalooza' was used during World War II since its tricky pronunciation could identify native, English-speaking Americans. But the sinister history of the word 'shibboleth' illustrates how friends and enemies are separated by fine, linguistic lines. Any stranger in a foreign land appreciates the vulnerability this entails, especially the fear of being outed as a foreigner and exposed in a hostile environment."[17] Here shibboleth stands for the displaced, the "out of place," and otherness.

Shibboleth is a prosaic word but sounds as though it should be the name of a monstrous creature. The name is adopted as such in entertainment and gaming.[18] It also crops up in online searches for academic papers. The word has caught on as a brand, notably in accessing academic texts online.[19] As with most words, the sound of the word contributes to its usage, nuance, and endurance. Pronounced "correctly," the word starts with the /sh/ sound that we use to instruct others to keep quiet. I will return to this thread in chapter 3.

Talking Trash

Ways of talking help define spaces. Like most urban commuters I have learned to tune out other people's mobile phone conversations. But when I'm forced to attend to a one-sided overloud conversation the interlocutor may as well be speaking in code: "She said that? . . . He did it then. . . . Ask him to give it to me when I'm there." *Deictic* utterances are those for which the listener needs to know the spatial and temporal context in order to establish meaning, and just hearing one side of a conversation makes it difficult to discover that context. It is in part a characteristic of grammar and how we use pronouns. Without the referent ("John," "the man in the seat opposite," "the boss"), "he" and "him" are insufficient to know who someone is talking about.

Competent language users are of course adept at speaking to one another in this coded way so that potential eavesdroppers would have to work hard to get the gist of the conversation. Avoiding proper names and other overt references to the subject matter is one way of "encrypting" the conversation. Another is to use abstruse terminology. In discussing this kind of coding I've drawn on themes developed in my 2010 book *The Tuning of Place*. There I focused on the mimetic and repetitive nature of arcane signals to claim space. Groups of friends would deploy vocal mimicry when they enter each other's company. In studying sound and the voice in the urban context, Jean-François Augoyard and Henry Torgue noted how teenagers' conversations are often filled with "onomatopoeia, interjection, and deictic words borrowed from the media or cartoons."[20] The listener needs to know the context in order to ascertain any meaning. Such exclusive uses of terms in language, whether spoken in groups or into mobile phones ensures that if the conversation is overheard it is not understood. That's one of many means of hiding messages in the cryptographic city.

The quote about teenagers is from Augoyard and Torgue's book *Sonic Experience: A Guide to Everyday Sounds*. It makes sense in the context of codes and the city to think about the everyday tactics deployed by ordinary language users to make themselves clear or obscure. Used out of context we might also describe such exchanges as vulgar, common, or even offensive. It is interesting that such banter is associated with the street. We draw on the terminology of urban infrastructure to identify this coded use of language ("street talk," "gutter language") as indexed by the aptly named online *Urban Dictionary* (urbandictionary.com).[21] Some languages and speech patterns have developed specifically to evade comprehension to those outside the circle of interlocutors. For example, Polari developed from slang among actors, circus people, prisoners, and dock workers. It is mostly associated with LGBTQ+ subculture of the 1950s.[22]

Many city features are hidden within code, or at least covert messaging. Navigation provides an obvious example. There's a famous New Yorker cartoon by James Stevenson of a policeman giving directions to a visitor from out of town. The thought balloon above the policeman shows a conventional map, with arrows clearly charting the route to the visitor's destination. That's what this knowledgeable local understands. But the thought balloon above the visitor shows a confused jumble of arbitrary map bits and a tangle of arrows. Navigational instructions require a degree of familiarity to be of use. To encounter a city for the first time is to encounter a series of coded messages. In time, seasoned residents can say, "I understand the code because I've been around it for a decade." These are tacit dimensions of language and social interaction. Under a coded system, what gets communicated is opaque to others in the room. Only the supposed recipient knows what the message is, or that there even is a message.

In these and other ways cities are colonized and formed by layers of messaging intended or customized each for their respective constituencies. But then messages get intercepted, relayed, retranslated, recoded, diverted, and distorted. I see those processes as the general communicative milieu to which cryptography belongs, or perhaps cryptography stands in for the whole communicative enterprise of the city. Before pursuing the ubiquity of cryptography there's more to be said about writing in the city.

Blocks

Any student of urban metaphors will notice the presence of *the block* in the city. There are city blocks. Buildings are made of blocks of stone. Freemasonry trades in the romance of construction with blocks of stone. In the ruins of ancient cities, it is the stone blocks that resist the ravages of time and persist. In imitation of architecture, playful infants make and demolish structures made out of plastic and wooden building blocks. It is well known that the architect Frank Lloyd Wright played with Froebel kindergarten blocks as a child.[23] The 3D virtual platform of the video game Minecraft, at minecraft.net, presents a world made of one-cubic-meter blocks. Even people, animals, and trees are made of blocks in this surprisingly immersive world. As if to illustrate my case for the cryptographic city, among Minecraft's major landmarks is a Masonic Lodge and the imposing neoclassical building known as The Uncensored Library, all made conspicuously of regularly sized blocks.

On the subject of text, graphic artists and page editors will *block* out the general layout of a poster or page with rectangles to indicate where they will place text and pictures, eventually to be printed. Early printing involved blocks of wood with carvings or elements in relief for stamping out patterns with colored inks and dies. Early printing presses used blocks with letters in relief for mass producing texts. We still talk about *blocks of text* as we write, type into a word processor, prepare presentation slides, and lay out pages. A block of text can be moved about while thinking less about the relationships between the block contents and the content of adjacent blocks.

Blocks are portable, at least in the practice and thinking of the designer or planner. More specifically a block is a bounded aggregation that can be treated as a single entity. When using drawing or 3D graphics software I will frequently group elements together so that they can be moved about, as a block.

A block is also an obstruction. A block in the drain is a single entity that needs to be shifted. A roadblock is a collection of people and paraphernalia whose effects on traffic flow are aggregated as a single block. It is the block as a whole that matters and exhibits the required identification and portability for that task and that moment. The block serves as convenient a descriptor of elements, processes, and arrangements in the city as it is in writing.

Spreadsheet City

The process of generating blocks of text is of course *writing*. To produce text is to write. I'm edging in this chapter toward a consideration of writing rather than text. As I will show, writing provides one of the many processes linking the city to cryptography. So, let us move toward *writing*, as an active term.[24]

What does it mean *to write a city*, implied in the title of this chapter? It is worth revisiting the importance of writing in cultural development. Doyen of media studies Marshall McLuhan encouraged digital pioneers with his popular 1960s accounts of how we are under the sway of the technologies of writing and print, and more recently electronic communications. Another scholar Walter Ong (1912–2003), a historian and student of McLuhan, reflected on the influences of technology in culture, not least the way technologies of writing and print have influenced cultural and intellectual developments—that is, the way societies think: their politics and philosophy. He wrote about the influence of bookkeeping and ledgers to be examined further in the context of cryptocurrencies in chapter 7.

Scholars such as Mary Poovey concur and emphasize the development of double-entry bookkeeping in basic commerce such as trading, banking, and lending money.[25] The double-entry aspect refers to keeping two ledgers: one for income, the other for expenditures. These scholars argue that the invention and popularity of double-entry bookkeeping not only aided commerce but also influenced the development of certain intellectual practices, not least, the way we think about order, rationality, and how we present and arrange information, and by implication knowledge.[26] Ubiquitous spreadsheet applications, task lists, schedulers, and online calendars provide contemporary illustrations of how wedded we are in business and personal organization to the tabulation of information.

For Ong, efficient bookkeeping set in train a way of thinking of things in the world as recorded and assigned a monetary value. "The arts and sciences could be viewed as a mass of 'wares,'"[27] he writes, "Here the most diverse products are mingled on an equal footing: wool, wax, incense, coal, iron, and jewels—although they have nothing in common except commercial value."[28] With this kind of calculative assessment, you don't need to look at the wide-ranging qualities of those items: "One has to know only the principles of accounting."[29]

Ong showed that the utility of bookkeeping, ledger writing, provided a benchmark for rationality. The ledger renders transactions transparent, able to be scrutinized, following a calculable logic. The ledger can also be reproduced, as can the procedure, to give the same result. To see the world through the lens of the accountant's ledger is to see a world conveniently divided into transactional components, classified, compared, weighed, evaluated, and eventually balanced.[30] These are further aspects of the city as writing, the city as ledger or spreadsheet, making way for a consideration of city processes that draw on the idea of the blockchain in chapter 7, and hence on cryptography.

Writing the City

You can write *on* a ledger, write *up* a ledger, and even just "write a ledger." Poovey describes the use of the ledger as a "rule-governed writing."[31] I would add that cities are similarly to be written about and written on. Of course, we not only write, but we also read. There is an attractive symmetry in the idea that you might both read and write the city.[32] To write something that isn't text, like a city, gives us pause. It operates figuratively—as a metaphor. To write a city suggests that the city is a book, song, poem, or ledger, or perhaps the second half (the vehicle) of the metaphor is open to whatever the idea of writing suggests to you. So, to write the city might mean to bring it into existence with something like the creative energy that goes into writing a song or a book.

I'm content to think that planners, designers, and citizens are authors who not only write reports and draw diagrams and plans, but who also write our cities. To bring in a performance metaphor, we city dwellers are also improv artists who make it up as we go along. We write and rewrite the script as we move. We write the city as we go about our daily business, all the more now as we (literally) write and read messages on our smartphones on the go, and leave digital traces, like pen strokes, as we negotiate the city.

Some joggers and cyclists carefully calculate routes through the city that result in drawings when viewed on a digital map application.[33] Such "GPS art" writes and inscribes invisible graffiti across the city: dinosaurs, skulls, portraits, logos, names. We write across the city by our movements anyway.

The density of our movements writes like a highlighter pen or sharpie indicating what people value and avoid spatially. Citizens also write over their cities by exerting influence, sometimes unwittingly, through the data they provide or through engagement and direct action via intermediaries such as designers, planners, developers, and decision makers. Citizens as city writers also offer up subversive texts, alternatives, and resistance to city transformations. We also write, annotate, mark up, and inscribe our presence across the city in other secretive ways to be discussed in subsequent chapters.

Writing and drawing are clearly related. The *-graphic* part of the word *cryptographic* makes us think of drawing, but the affix has most to do with writing. The etymonoline.com website describes the common abstract noun ending *-graphy* as a "word-forming element meaning 'process of writing or recording' or 'a writing, recording, or description.'" So, we have geo*graphy*, biblio*graphy*, choreo*graphy*, crypto*graphy*, which primarily write and therefore describe the earth, books, dance, and messages. "To draw" and "to write" are similar in that they can both function as either transitive or intransitive verbs, i.e. they function with or without an object: She draws; she draws a tree; he writes; he writes a book. You can draw just about anything, but in normal usage, writing is more limited. The songwriter Arthur Hamilton exercised simple verb "misuse" with the titles "I Can Sing a Rainbow" and "Cry Me a River." I say "misuse" as that's one of the ways philosophers and poets have characterized metaphor, as a deliberate misclassification: putting rainbows in the same category as songs by suggesting they can be sung, and rivers in the same class as tears and emotions as if the result of sadness or melancholy.

Here is an example of our readiness to think we can write physical objects like cities into existence. The early computer game series *Myst* (1993) played with the grammatical asymmetry of reading and writing. The protagonists had somehow acquired the art of "writing ages," which are islands linked together by specially written link books. The games are populated with 3D models and renderings of tangible spaces, and virtual handwritten books containing descriptions and diagrams of imagined mechanisms and places. Players don't need to look for consistency or logic here. Who would not want to write things into existence with the stroke of a pen! Metaphor and imagination accomplish that for us.

Printing the City

Drawing and writing involve similar processes as long as we think of placing a pen, brush or quill in contact with paper. In the history of technology print supplanted writing by hand as a means of preserving texts. Printing involves a set of procedures that involve reproduction via mechanical processes. How does the transformation from writing to print impact the evolution of cities?

To fill a ledger with numerical data was a particular method of writing. By Ong's account, the invention and adoption of orderly double-entry bookkeeping was among a series of events on the trajectory to modernity. Scholars think of Johannes Gutenberg's (1400–1468) invention of the movable-type printing press as a pivotal moment in the cultural and social development of Europe and beyond. By Marshall McLuhan's reading the printing press ushered in a revolution in thought.[34] Printing firms were able to deploy individual, durable typographic elements (letters and punctuation marks) manufactured in metal and arranged in rows to produce a page of text, the inked imprint of which was transferred to sheets of paper, repeatedly. For McLuhan the industry of movable type was the final triumph of visual over aural culture helping us humans see the physical world as distinct from ourselves, as able to be inventoried, documented, and studied. For McLuhan, the ability to print and distribute multiple copies of texts cheaply and quickly reinforced "homogeneity, uniformity, and continuity"[35] in communication and in culture.

There are other aspects to the technique of movable type. Scholars have considered the Gutenberg revolution in terms of the tools of printing—the characteristics of movable type itself, which after all promoted the utility of modularization and rearrangement. Typographers in printing offices would arrange and rearrange typographic symbols. They could also substitute one symbol for another. Movable type anticipated the intoxicating freedom we have now with on-screen editing, as I move text around, correct, and substitute one symbol, word, phrase line or paragraph for another.

Alberti's Cipher Wheel

The power of moving text across a page predates the printing press. It is worth recalling the earlier innovation of Ramon Llull (1232–1316), the

Catalan polymath who invented the so-called "Ars brevis," "Llullian wheel" or "memory wheel." I first encountered this invention, or at least its representation, as a moment in the lore of artificial intelligence (AI). By the AI account this disk-shaped paper Llullian wheel served as a rudimentary logic machine that worked with combinations and calculations. Scholars are keen to position this device as an early progenitor to the computer.[36] The wheel consisted of a series of concentric disks marked out with godly attributes or "dignities." Historian of cryptography Quinn DuPont identifies the attributes: "goodness, greatness, eternity, power, wisdom, will, virtue, truth, and glory."[37] The wheel would allow a scholar to combine these attributes in various ways to demonstrate the logic of an argument, notably to explain Christian doctrine. According to DuPont: "Each 'dignity' could be combined according to particular rules, which amounted to a method for investigating reality. That is, this method was a way to do work and to actively investigate or 'compute' the world."[38]

Llull's pre-Gutenberg logic machine operated as if ideas can be manipulated, combined, and moved about. The Lullian wheel was also a memory device, a means of recalling and copying the ideal order of the world. According to DuPont, "Lull reconfigured the theory of representation that previously relied on the complex web of resemblances, as it had been handed down to him through ancient and medieval transmission."[39] The *Stanford Encyclopedia*'s entry for Ramon Llull illustrates the wheel, its variants, tables, and complicated and occult nomenclature.[40] Even a cursory glance shows how wedded this kind of rationality is to a mimetic view of the universe, as if representing or imitating higher realities.

Intellectual life pre-Gutenberg appears alien to our current age. For example, Ramon Llull's memory wheel cannot easily be inserted into a linear history in the development of the computer. The device has more in common with secret societies than with computation. A Spanish Freemasonry lodge bears Llull's name. As with the Medieval arts in general, this particular calculative invention confounds attempts to distil any purely utilitarian and computational operations from this occult, mysterious, and cosmological apparatus of reproduction—of divine mimetics. The memory wheel operated in a world imbued with an understanding that art, knowledge, and beauty derived from imitation. The Llullian wheel serves nonetheless as precursor to the cipher wheel, the invention of which is attributed to the influential architect of the Renaissance Leon Battista Alberti (1404–1472).

According to DuPont, "Lull's development of an active system using rotating wheels, with its particular history of representation, was an important precedent for Alberti's invention of the cipher wheel."[41] Alberti's cipher wheel has similarities, but departs from Llull's memory wheel in significant ways, not only in its purpose but also its format. The cipher wheel is more consistent with the combinatorial instrumentality of the Gutenberg revolution.

Gutenberg's printing press allowed for the accurate preservation of texts, contributing to the idea that knowledge accumulates. But there were other benefits as it released energies from the laborious task of copying and transcribing texts by hand and other cumbersome and unreliable processes for reproduction such as woodblock printing. According to social geographer and historian David Lowenthal, "Energies released from tasks of retrieval and preservation could focus on other creative activities, thereby detaching inspiration from the bondage of imitation."[42] So, contrary to Llull, Alberti's cipher wheel, developed under the sway of the printing press and movable type, and signifies a transition to something resembling the modern era. As did Llull's wheel, it adopts the grammar of the sacred circle and axis, and is divided into "houses," as if they are constellations, but it also adopts the new functional processes of mass production: reproduction, modularization, and combination. Like the printing press, Alberti's cipher wheel combines letters of the alphabet. For DuPont, "Both the movable type press and the cipher wheel utilized reproducible, modular, indexical, and combinatory forms of representation."[43]

Alberti's cipher wheel was for cryptographic communication. The technique enabled the transmission of confidential messages between leaders—without the need for interpreters. Alberti wrote, "I can justly consider this cipher worthy of sovereigns, who can use it quite easily, with little effort and without being encumbered by use of an interpreter."[44] Those in communication may be many miles apart even though they share the same language. The interpreter was likely a functionary who would receive a coded message and had the equipment and the knowledge to decode it and pass the translation on to the sovereign. Couriers and interpreters lived dangerous lives, not least for the secrets they knew. Human translators can't easily forget, and easy-to-use translation machines reduced the risk of message leakage from a circuit of human intermediaries.

Alberti Code

I was first alerted to Alberti's cipher wheel in a history display in the Berlin Spy Museum. My own training in architecture had failed to register Alberti's expertise in cryptography. Alberti introduced the cipher wheel in his essay *De componendis cifris* (1466).[45] The cipher wheel is a simple device with an inner and outer disk cut out of bronze sheets (or paper), each marked into twenty-four divisions. The outer disk lays out the letters of the Italian alphabet in upper case and some numbers. The inner disk has lowercase versions of these symbols but in an arbitrary order. The two disks can be rotated freely about the same center. The device provides a simple way of mapping the letters from a text message to create a coded version. As long as the receiver has the same device and knows the alignment of the disks, it is simple to decode the message.

Unfortunately, it's also a simple matter for any codebreaker (cryptana-lyst) to work through various letter combinations to come up with a good guess at the correspondences, and eventually the hidden message. Certain letters occur more frequently in any language and that acts as an initial clue for the codebreaker. The clever idea of the cipher wheel is that the coder shifts the alignment of the inner and outer disk at certain points in the message and indicates these shifts to a new alignment by inserting a special character in the coded message. This new mapping effectively scrambles any one-to-one correspondence between a symbol and its coded equivalent, making it harder for codebreakers.

Alberti's cipher wheel looked something like the drawing in figure 2.3, providing a simple equivalence mapping between letters of the alphabet. He used the Latin alphabet, and excluded a few letters as well, perhaps for symmetry. According to a helpful YouTube clip by Ciphertown,[46] the cipher disk user would represent H with two Fs, J with two Is or similar secondary coding. Here's a demonstration of the method. I'll restrict myself to Alberti's character set here to encode the simple communication: WRITE ME A CITY. With the disks in the position shown in figure 2.3 that would result in a simple one-to-one mapping between characters. As long as you know how to align the two disks you could convert my secret message to hmvip &p g lvih. That's actually "VRITE ME A CITV," a necessary variation due to the reduced character set. With the same cipher disk design and alignment,

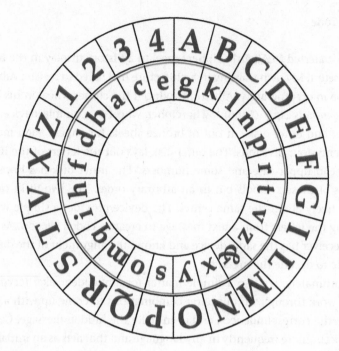

Figure 2.3
Alberti's cipher wheel. Drawn by the author.

the receiver could decode that back to the original message. But that's also relatively easy for an intercepting codebreaker to decipher, even without a similar cipher disk, especially with a longer message, considering letter frequency, something about the context, and iteration through combinations.

Here's another encryption that is more difficult to decode using the same disk design and knowing that the letter g is to serve as an indicator: Ahmvip &p Cc grmq. The appearance of the letter A means align g on the inner disk with A on the outer disk. Taking each letter at a time, when intended message recipients encounter the uppercase letter C they would rotate the inner disk to align g with C to give a different set of mappings (figure 2.4). The coder can insert such rotation cues throughout a long message. That would make it extremely difficult for an intercepting codebreaker to decode the message.

It is a simple process, and there are elaborations on the method to make code breaking even more difficult. Alberti described the shift in alignment as "changing the index", and the family of such methods of shifting the

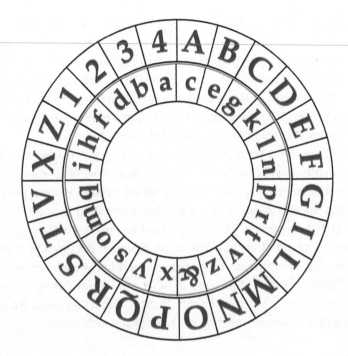

Figure 2.4
Cipher wheel in the rotated position. Drawn by the author.

alphabetical ordering at different points in the message is now called *polyal-phabetic encryption*. In *The Code Book: The Science of Secrecy from Ancient Egypt to Quantum Cryptography* Simon Singh provides a helpful history of the cipher wheel's use and elaboration since Alberti.[47] He speculates that such a system could easily have deployed a key in the form of a sequence of let-ters, perhaps a key word known only to the sender and receiver that would indicate at which letters the alignment of the discs should be adjusted.[48] Though much more elaborate, the German Enigma machine, the electro-mechanical coding device invented by Arthur Scherbius at the end of World War I, belonged to the same family of encryption machines; this device and method eventually were cracked by the codebreakers of Bletchley Park.[49]

Alberti's interest in mathematics and geometry led him to other cod-ing techniques as well. According to the reading of mathematician and architectural theorist Lionel March, Alberti also projected cabbalist and covert occult methods into the geometry and tiling of the facade for the church of Santa Maria Novella in Florence in such a way that they would go undetected by church officials.[50] The medium was not of symbols of the kind used in Freemasonry, but "arithmogrammatic 'calculations' to demonstrate theological, particularly christological truths."[51] March identifies Alberti's own name in the coding, as well as "Yahweh." March admits his detailed analysis of the geometry is speculative, but it's fair to say that Alberti's interest in cryptography was integrated into some of his architecture, notably the facade treatment of Santa Maria Novella.

Alberti was in the company of other renaissance polymaths who applied their inventiveness across architecture, sculpture, writing, music, astronomy, and other arts and sciences. That Alberti dabbled in encryption provides just another example of renaissance innovation at work. But by a more interesting reading, Alberti's penchant for encryption is integral to his understanding of architecture. This is the argument advanced by DuPont.[52] It seems that the common factor between architecture and encryption is the influence of the printing press, in other words, movable type, invented and adopted during Alberti's lifetime. Alberti's innovation paralleled radical changes in how people understood language and architecture.

Portable Type

Earlier in this chapter I referred to the portability of text. Portability implies movement, and the dynamic aspects of letters and characters in the printing process as suggesting arrangement and rearrangement. Movable type exerts an influence on architecture and the city. There's a long tradition that considers architecture as the art of arranging things—according to Vitruvius, "the putting of things in their proper places."[53] In his book *Architecture in the Age of Printing*, architectural theorist Mario Carpo maintains that architecture as an art of arrangement reached its apogee with the invention of the movable-type printing press.[54] Alberti was impressed by the idea that letters could be selected and arranged to produce reproducible printed pages. Alberti admitted this influence, and in his book *Art of Building* even used letters of the alphabet to describe the profiles of cornices and moldings. For example, the astragal is shaped "like the letter C surmounted by the letter L."[55] That use of letters hinted at the influence individuated letters had as part of the cultural vocabulary of the time, as metaphors and explanatory devices. It also suggested the "misuse" of letters, a willful obsession with their shape and rearrangement to produce something outside of their usual signifying function, a rudimentary "word art."

Other than this creative digression the architectural tradition of the time seemed to lack a language of innovation. Like the letters of the alphabet, the elements of architecture (columns, walls, roofs, rooms) did not need to be invented, but selected and arranged correctly. There was no appeal then to creation or innovation in the parts. Nor was innovation encouraged in the way elements were to be arranged. In fact, architects were warned against inappropriate placement. Alberti wrote about aberrant forms: "When even the smallest parts of a building are set in their proper place, they add charm; but when positioned somewhere strange, ignoble, or inappropriate, they will be devalued if elegant, ruined if they are anything else."[56] Nature provided the model for perfect arrangements.[57]

By a twenty-first-century interpretation, it's as if Alberti's negative examples harbored the seeds of dissent, if not invention. Aberrant combinations and rule breaking, as in architecture and all the arts, did of course exist in these traditions, but was exercised in the outcast world of the carnival and the trickster.[58] Though *The Art of Building* was saturated with convention and propriety, I'm prepared to concede that Alberti was on the way to thinking of architecture in terms of inventive arrangement and rearrangement. By my reading, that observation aligns architectural innovation with the affordances of the printing press and with media theorists such as Marshall McLuhan.

Leibniz's Secret Machine

The flowering of the intellectual development in Europe known as the Baroque period followed two centuries after the introduction of the movable-type printing press. Chief among the period's luminaries was the mathematician and philosopher Gottfried Leibniz (1646–1716). There are at least three touchpoints connecting Leibniz with cryptography. First, Leibniz invented and built a mechanical calculator. But less well known is his unbuilt model of a machine for encrypting and decrypting messages, using the polyalphabetical encryption method, though with many more moving parts than Alberti's cipher disc. In his account of Leibniz's machines, the philosopher Nicholas Rescher notes: "The calculating machine performs mind-like processes in relation to reasoning, and the cipher machine performs mind-like processes in relation to communication."[59] So, Leibniz aimed to demonstrate by his mechanical inventiveness the two main functions of human cognition as he saw them: *calculation* and *communication*.

Leibniz pitched the cryptographic machine to his emperor, who failed to provide support for the manufacture of a prototype. As Leibniz thought the device was only for princes, according to Rescher, "this apparatus was Leibniz's most closely guarded secret."[60] It was never built, though others (e.g., Nicholas Rescher) have since attempted to replicate it.

Second, much of what we know about Leibniz's life comes from his biographer Johann Georg von Eckhart (1664–1730). Around age twenty, Leibniz spent a year as the secretary of a secret society of alchemists. An article by George Ross probes the claims and counterclaims of Leibniz's commitment to alchemy. Ross provides an interesting excerpt from Eckhart, delivering a tip on how to inveigle your way into a secret society:

> "Now, since he [Leibniz] was curious about everything, he was very keen to have some practice in chemistry as well; so he considered all the various ways of getting access to these secrets. The director of the society was a priest. So he devised the following trick, as he himself often told me with a laugh. He got hold of some very difficult books on chemistry, and noted down the obscurest phraseology he could find in them. Out of these he composed a letter to the priest, which even he himself did not understand, and in it he also requested admission to the secret society. On reading the letter, the priest came to the conclusion that the young Leibniz must be a true adept, and not only gave him access to the laboratory, but asked him to become their assistant and secretary."[61]

Ross is keen to dispel any notion that alchemical mystery cultism had significant impact on Leibniz's mature philosophy, apart from demonstrating his ambition and opportunism. I'm prepared to see this episode as evidence of a philosopher's curiosity. Curiosity draws people to secrets, and even to promote and exploit them. Like many luminaries of this period, Leibniz knew of occult practices, and was even familiar with Gaffarel's writing on "the celestial Hebrew alphabet."[62]

Third, Leibniz famously defined, or redefined the concept of the monad. According to the *Stanford Encyclopedia of Philosophy*, "The ultimate expression of Leibniz's view comes in his celebrated theory of monads, in which the only beings that will count as genuine substances and hence be considered real are mind-like simple substances endowed with perception and appetite."[63] In other words, the world is made up of perceptual units. According to Leibniz's philosophical idealism, monads are the only things that are *real*. Monads are also indivisible. Among the definitions he provides in the essay *Monadology*, Leibniz asserts, "the Monads have no windows, through which anything could come in or go out."[64]

Leibniz's *Monadology* is of interest, especially as inflected through the writing of the twentieth-century philosophers Gilles Deleuze and Felix Guattari, for whom the term *monadology* invokes the notion of *nomad*ology, the study of how sedentary power is permeated by itinerants as if from the outside. The simple formulation of the "windowless" monad operates at least as a provocative metaphor. Through its explanation in terms of spaces and windows, nomadology brings Leibniz's philosophy further into the orbit of architecture and secret places, such as *crypts*. In his book *The Fold: Leibniz and the Baroque*, Gilles Deleuze states, "A 'cryptography' is needed which would both enumerate nature and decipher the soul, see into the coils of matter and read in the folds of the soul."[65] Deleuze elaborates on the metaphor of the fold to explain the art and architecture of the Baroque, a useful subject with which to culminate this chapter on writing, arrangement, and architecture.

Fabrics and Folds

Folds appear in fabrics. What is the fabric in this case? It is a kind of surface, boundary, or border between two places or conditions. Put simply, the fabric to which Deleuze refers is the boundary within the Platonic universe between the world of ideas and the material world: you could say, between the macrocosm and the microcosm; the heavenly and the earthly; or souls and materials. For Plato that would be a boundary between the real but invisible and immutable world of ideals, of which the material earthly world is a pale shadow.[66]

I began to appreciate the significance of the fold in Baroque architecture on a visit to St. Peters Cathedral in Trier, Germany. The cathedral is mostly Romanesque, but behind the main altar the visitor has a view into the Baroque chapel for the relic of the Seamless Robe of Jesus. As illustrated in the Trier chapel, Baroque surfaces do fold, as if affording glimpses between realms. Deleuze provides an account of the Baroque fold that has inspired many speculative architectural and urban design studio projects.[67] Notice Deleuze's reference to levels, compartments, and windows: "There are souls below—animal, open to sensation—or even bottom levels in souls, and the coils of matter surround them, envelop them. When we discover that souls can have no windows to the outside, we will need, at least at first, to think of this in reference to the souls above, the rational souls, which have risen to the other level ('elevation'). It is the upper level which has no window: a darkened compartment or study, furnished only with a stretched cloth 'diversified by folds,' like the bottom layer of skin exposed."[68]

He supports this description with a crudely drawn cross section through an allegorical building he calls "The Baroque House." There's clearly an upper and lower story, and the folding elements "spiral" into the upper reaches. This is an inverted crypt—there are no windows in the upper story. In this book (*The Fold*), Deleuze references Leibniz's concept of the window-less monad, the indivisible unit of perception, or the "mind-like simple substances endowed with perception and appetite" as explained in the *Stanford Encyclopedia*.[69] These are further reminders of the confluence between architecture and cryptography.

Of even greater relevance to the theme of cryptography, Leibniz was a distinguished mathematician (as well as designer of a cryptographic machine), responsible in part for the invention of calculus, which deals in rates of change, in other words, the slopes of mathematical curves, as well as areas under curves, techniques that will assume greater significance as I probe methods of digital encryption.

In this chapter I examined writing and arranging letters as a simulacrum for arranging city elements (public and private spaces, buildings, parks, roads, street furniture, signage), a process the theories of which are inspired by ordering and reordering letters in a movable type printer. That took me on a journey through Alberti's cipher wheel, one of the earliest systematic and replicable methods for encrypting messages. The technology later developed into sophisticated coding methods such as the polyalphabetic Enigma coding device. Alberti's cipher wheel and his influence in architecture and urbanism entitles architecture to claim a pivotal place in the history and theory of cryptography, and hence as curator of the cryptographic city. I also recruited the rationalism of Leibniz, the Baroque, and Deleuze's provocative spatialization of the monad as progenitors of the cryptographic city idea.

In order to link the spatial *assemblies* or *combinations* of the city with cryptography, we could do well to examine cryptographic puzzles, the subject of chapter 4. In the meantime, if cities are places of reading and writing, they are also sites of coded communication, the subjects of chapter 3.

3   Place Is the Code

There are many ways to make a place secure, including combination locks on security doors. The sight of a touch pad is enough to deter most people from trying the door handle. Spaces that signal their security measures may also signal the converse: something valuable is concealed here; it is worth breaking in. The discipline that foregrounds signals, signs, and messaging is known as *semiotics*. As communication is central to the cryptographic city, I want to start this chapter by positioning the securities afforded by urban cryptography within the wider field of semiotics.

Signs in the City

Semiotics focuses on the concept of the sign, the basic unit of communication whether spoken, written, implied, or embodied in architectural and urban objects. Semiotics is a practical matter that touches on how things communicate but also how they function in physical and social contexts. Semiotics assumes the all-pervasive nature of communication within human, biological, and even inanimate systems.[1] Winfried Nöth's expansive *Handbook of Semiotics* illustrates the broad scope of the field and the wide divergence of views it encompasses and that it claims within its orbit.[2] I'm thinking here of language philosophy, structuralism, poststructuralism, and deconstruction, with Wittgenstein (see chapter 1) exerting considerable influence across each. These philosophies are widely applicable across architecture and urbanism as I've recounted in several of my own publications.[3]

I presented James Gibson's theory of *affordance* in the introduction. Though framed in the language of ecology, Gibson's ideas find a home in the field of semiotics.[4] The way I read Gibson, the world tells us how it is to be understood, interpreted, accommodated, cared for, and used. Touch pads, steps, doors, letterboxes, buildings, roads, and all the elements of the city afford actions, such as entering, walking, posting, dwelling, and commuting. That such objects carry affordances elides with the semiotic insight that they also *communicate* those possibilities. The extent to which a door makes clear whether you should push or pull to open it depends both on what actions it supports and what possibilities it signals.

As Nöth evidences in the *Handbook of Semiotics*, the discipline of semiotics also claims cryptography in its lineage. Semiotics includes the transmission of secret messages, secret access codes, the means of securing data and the whole apparatus of cryptography. In what follows I want to show that prosaic everyday semiotic practices in the city are subject to processes of coding and decoding information found in cryptography.

Any communication takes place within a context of multiple signals across many communication events. Any particular message lies within a much wider context of messages and communication practices. According to one of the standard models of communication, a sender codes their thoughts into some communicative medium and passes them through a conduit to be decoded by a receiver.[5] But in the case of everyday language, if communication involves coding and decoding then the coding system is also "transmitted" through the message corpus that includes previously shared exchanges. The means of decoding the message is transmitted along with the message, and in the company of other messages exchanged within a language community. We learn and transmit the code as we communicate.[6]

It serves well the design of the urban environment to think that the code is potentially part of the message, that communication events include the means of their decoding, or we could say of interpretation. Consider everyday tools, such as a hammer or toothbrush. Tools that are well suited to the task at hand, whether objects or words, make clear how they are to be used. They are their own codebook. Well-designed door handles signal to us by their shape, materials, and positioning whether they are to be pushed, pulled, or turned.[7] The current design of supermarket self-checkouts provides a negative example. Should I put my purchases to the right or left of the scanner? Where do I put my own recycled shopping bags? The current design of such machines rarely makes obvious how they are to be used, without lists of instructions, audio alerts, and the intervention of shop assistants. In less-managed contexts, physical signage tends to accumulate. I think of paper notices adhered to the wall adjacent to our office printer, the dos and don'ts of its use and maintenance, some of which contradict. In his book *Site Planning*, urban planner and theorist Kevin Lynch expresses misgivings about notices as "ugly and chaotic, not by their nature, but because they are thoughtlessly used, ambiguous, redundant, and fiercely competitive."[8]

Applying this code metaphor to space and place: the code that "unlocks" the secrets of a place are in the place. What does such a place look like? There's nothing remarkable about the idea of a place that contains within it the means of its translation, decoding, and interpretation. A place that is "legible" will contain tools, objects, and affordances that enable us to do useful things, which is to say that enable us to *understand* that place.

Formal signs as printed texts, notices, or instructions assume the role of rules, as if handed down or approved via some authority structure. Notices as textual signs populate the urban environment and tell us what to do and what not to do: "Place goods in bagging area." Unnecessary urban rules come across as restrictive, censorious, and authoritarian. Places, buildings, and devices that present to us as a plethora of printed, vocalized, and displayed instructions have failed in some important design sense.[9] By way of contrast, places that enable self-evident social practices suggest nuance and sensitivity to the user and to context. Such place-based unwritten practices are *norms*. Designs that rely on text messages, signs, and audio that tell us what to do displace norms with rules.

Urban Affordances in Context

Citizens learn to interpret, apply, and perform actions in social contexts. I soon learned how to use the unsympathetic self-checkout machine. I watched others. The machine and staff provided feedback when I did something contrary to the machine's function, such as placing my own shopping bag on the purchased goods side.

Norms are baked into well-designed and evolving urban places. I say "place" here. There's a social dimension to concepts of place. We tend to emulate how others use a place. Sometimes we just need a nudge to alert us to the norm, as when a pedestrian berates a cyclist for riding on the sidewalk. Bike parking racks do not need a notice that says, "Park your bike

here." The sturdy iron frames secured to the pavement appear to be emi-
nently practical as a safe place to chain your bike. We don't need notices to
tell people to gather in large central city plazas. The location of the place in
relation to circulation routes, the control of vehicular flows, the welcoming
nature of the architecture, the presence of retail and entertainment, and its
accommodation of properties of bodies and movements abet people's will-
ingness to congregate. Increasingly, the design of places takes into account
their potential to attract, even as settings for digital photography.[10] Curbs
on the edges of roads make it clear that cars must not encroach on side-
walks. Low, smooth, self-cleaning walls invite people to sit.

Urban designers and theorists such as Jan Gehl provide many examples
of successes and failures in the arrangements within public urban space
attributable in many cases to the communicative and functional aspects of
places.[11] Places that are effective signal to you how you are to use them, or
at least they provide an indication of the scope of their use. They accommodate social norms, unwritten understandings about what constitutes
normalcy in this context.[12]

The city is permeated by written and unwritten codes. Admittedly, that's
different from the idea of the city as a medium in which people communicate from sender to designated recipient using overt cryptographic methods
and devices: Enigma's polyalphabetic electromechanical coding, Wittgenstein's substitution cipher or digital encryption. But I contend that even
without such overtly cryptographic operations the city is a place that deals
in the affordances of cryptographic processes and cultural outlook. Places
participate in affordances that reveal, conceal, combine, suggest pathways
to follow, hark back to original sources, facilitate interactions, and engender trust. Place is the code. The code is in the place.

Illegible Cities

One of the affordances of cryptography is to conceal plain text by rendering
it illegible to most but legible to a few. Legibility is a major dimension of
city spaces. In his guidelines on planning and design in cities, Kevin Lynch
outlines the means by which city places are enlivened by spatial elements:
transitions, views, sequences, forms, textures, details, contrasts, orderings,
and diversities that color our expectations within spaces. Key among these
attributes is *legibility*. The various parts of the urban environment should

be so structured that people "can relate them to each other and can understand their pattern in time and space."[13] He asserts that "the general framework of a living space, as well as the linkage of its public places, must be
legible—not only in the street but also in memory."[14] This is a practical
matter and aids in wayfinding and talking about a place. But legibility "can
be a source of emotional security and a basis for a sense of self-identity and
of relation to society."[15] He also suggests that legibility contributes to social
cohesion.

A legible city exhibits flexibility in how it is presented as maps, sequences,
and schemas.[16] As well as seasoned inhabitants, tourists should find the city
legible, whatever the varieties of urban experience.[17] He argues for a temporal legibility as well: "The sensuous environment may be used to orient
its inhabitants to the past, to the present with its cyclical rhythms, and
even to the future, with its hopes and dangers."[18] Urban theorists generally
alight on historic cities to illustrate urban legibility: the Piazza San Marco in
Venice,[19] the Bogro Allegri, Florence,[20] the Georgian public spaces in Bath,
England.[21] Such places are time-honored, approved by scholars, and popularized by countless tourist guides, travel documentaries, websites, and
Airbnb hosting sites.[22]

In *The Image of the City*, Lynch argues for legibility while recognizing the
value of its converse—illegibility: "It must be granted that there is some
value in mystification, labyrinth, or surprise in the environment. Many of
us enjoy the House of Mirrors, and there is a certain charm in the crooked
streets of Boston."[23] However, he qualifies the value of such surprise by stat-
ing that it must occur within an overall framework that is legible, and "the
labyrinth or mystery must in itself have some form that can be explored
and in time be apprehended."[24] Lynch suggests the need for legibility is
not a universal requirement, that it is sometimes desirable that parts of the
environment are "hidden, mysterious, or ambiguous."[25]

I recall as a student visiting the city of Venice and wanting to get lost
in the tangle of canals and streets, to feel as though the city had indefinite
extent. But I still wanted to find my way back to the train station. We experience cities in different modes: as a tourist wanting to get lost, as a visitor
trying to find the way to a particular landmark or back to their hotel. For
some local citizens the city scarcely impinges on their awareness. These
are habituated residents, service personnel, postal and food delivery drivers and cyclists who need to cover as many bases within an allotted time.

Places designed and developed to provide their own code, the means of their own decryption, pose challenges appropriate to the mode of urban engagement. I'm prepared to release the city from the necessity to be always legible at all times and to everyone—hence my advocacy of the city as cryptographic.

Urban Contest

Norms imply constancy and consistency among the users of public spaces. Not everyone agrees on what is to happen in a place. Baked-in, place-based norms are often dictated by those with an investment in the status quo. Certain spatial practices emerge that run counter to the authorized and dominant semiotics of a place. Traditionally this would have been the carnival, pagan festivals, and street entertainments. Think now of flash mobs, pop-up markets, parkour, skateboarding, protests, street performances, outdoor sleeping, trespass, and other grassroots urban practices not yet appropriated or noticed by city authorities, urban researchers, mass media, and commercial interests. Such spatial practices are varied responses to the semiotics of place. What to one person is a place to sit is for another a place to spend the night. A traffic barrier becomes a climbing frame. A window ledge becomes a foothold for a climber. A metal balustrade on a bridge is a place to attach a love lock. Marginal and unsanctioned spatial practices reveal aspects of city fabric in ways that are hidden to the rest of us. Groups may have different responses to places and that delivers conflicts of expectations and practices.

Elements within the city also form as a response to conflict. Most city ordinances and regulations point to physical adaptations that were established in the wake of a disaster, including fire escape stairs, nonflammable cladding, shielded electrical cables, and drains that prevent flooding and disease. Lockable external doors have arisen or coevolved with developments in city fabric and with spatial practices to prevent the inconvenience of a conflict between public and private. Cities engage in competition across numerous fronts, not least between those who wish to protect, and those who want to break in.

Urban theorist Wendy Pullan has indicated that there really are no rules for urban conflict, productive or otherwise. Identification of the protagonists, their differences, and causes of conflict are fluid, contingent, and subject to the workings of interpretation, and rightly exercised, debated, and worked out in public life. So, an architecture that provides space for public life is crucial for the working of productive conflict, which she refers to by the Greek-origin *agon* meaning struggle, contest: "Place, by being structured in everyday activities rather than regulatory systems, can begin to open a territory where the necessary flexibility of *agon* can exist, with all of its paradoxes and ambiguities."[26]

By my reading cryptography captures part of that contest. At its most benign, cryptography presents as a puzzle or a game. More significantly, and traditionally, cryptography was deployed in war settings, as combatants would exchange information in their military maneuvers. Cryptography is *agonistic* across several dimensions, not least the contest between cryptographers and cryptanalysis, the coders and codebreakers, those who dispatch coded messages and those who seek to intercept or scramble those messages.

Cybersecurity in the cryptographic city intensifies a discourse of conflict, contest, and war. Hugo Hoffman's *Cybersecurity Bible* enjoins us to "think like the enemy"[27] and makes frequent appeals to defend against attackers. We should always assume there are active agents intent on sabotage. Concepts of cybersecurity instill anxieties about risks and threats. This adversarial framing of place is familiar enough to architecture, which includes castles, city walls, and other defenses in its repertoire. The early architectural theorist Vitruvius (ca. 80–15BC) paid substantial attention to city defenses, as well as catapults, ballistae, and other means of breaching city walls.[28]

This militaristic framing also underpins characterizations of the security challenge. In chapter 6 I'll introduce encryption methods. Three named fictional characters recur in cryptographic textbooks to help explain the methods. They are named benignly Alice, Bob, and Eve. According to Simon Singh in *The Code Book*, "Alice wants to send a message to Bob, or vice versa, and Eve is trying to eavesdrop."[29] That simple three-way drama implies persistent and active agency, malevolence from the third party (Eve) who wants to intercept communications between legitimate correspondents. It is a useful model in understanding the workings of cryptography, but it's a framing at odds with everyday life. In familial and friendly social settings, malicious overhearing, eavesdropping, and the interception of private communications is the exception rather than the rule. So it is difficult for many of us to tune in to the supposedly urgent demands of cybersecurity.

In friendly social settings, nonadversarial metaphors that depend on norms of trust apply. First is the *dilution* metaphor. A container holds something of value, our information. Secret keepers don't want that value diluted as it leaks and dissipates into general discourse. Second, think also of the value we attach to secrets. Secrets are objects of value to many and are to be kept, shared, and gifted as such. Third is an affective framing. Emotional states associated with pride, shame, and scandal feature in everyday relationships and guard against unwarranted interception of private communications. In contrast, cybersecurity aims for severe, inflexible, and all-encompassing protections to information, though I will show subsequently how cryptography is recruited in discourses about trust.

Some years ago colleagues and I penned an article we titled "Permeable Portals"[30] in which we advocated for the design of websites that resist firewalls and security protocols regulated via secure logins. As our students developed skills in interaction design, we would help them to resist the instinct to set up registration and login protocols to protect all aspect of data flows. Now, public-facing platform design (such as social media and gaming platforms) typically afford graded access to resources via "guest" access, levels of membership, and "freemium" permission structures set by default or in a user's profile. Not all users of a platform have to type personal details into a registration form or negotiate paywalls. Not all data and interactions need shielding all the time and for all time. These are urban concerns as well. Cities have long dealt with the physical arrangement of properties, zones, and districts and the permeability of their boundaries. Contest and the management of border conditions runs deep within information flows, as well as in design and the city.

Secure Places

Cryptography affords controlled access and security. To reveal and to conceal in urban settings is to negotiate matters of trust, risk, and security. Norms impact on security in urban settings via a kind of collective surveillance. "Everyone has his eye on you,"[31] wrote Thomas More (1478–1535). This is something akin to libertarian civil society. A sense of community and shared social responsibility promote security and appropriate behavior that can be abetted by environments and systems in which people's activities are in view rather than hidden away.

As an illustration of collective security and trust, the urban critic Jane Jacobs attributes the keeping of "public peace" not to the presence of police, but "by an intricate, almost unconscious, network of voluntary controls and standards among the people themselves, and enforced by the people themselves."[32] In contemporary spatial terms the contrast is most acute if we compare gated communities that lock in monocultural enclaves with open and diverse neighborhoods. Gates and locks advertise there's something to be protected. In vibrant, diverse neighborhoods with permeable boundaries people are *watching out for each other*.

Many studies indicate there is a positive relationship between busy places and security, or at least a sense of security. Security expert Bruce Schneier provides such advice in his book *Beyond Fear: Thinking Sensibly About Security in an Uncertain World*. The best way to feel safe is after all to "live in nice neighbourhoods where people watch out for each other."[33] When we think of how to be safe the first militaristic impulse is to build a fortress: "If someone is living in fear—whether it's fear of the burglar on your block or the fanatical dictator half a planet away—it's because she doesn't understand how the game of security is played. She longs for a fortress."[34] He advances the case that there is no condition of absolute security, in spite of the suggestion from news media and fiction, feeling secure is a subtle condition depending on context, and conditions and contexts change over time.

Feeling safe involves trade-offs. You need to be smart about your environment: "The smart way to be scared is to be streetwise."[35] There is scope in lived urban environments to communicate and implement measures that obviate the need for high walls, security fences, surveillance cameras, and keypad-operated checkpoints. Online communications and data transfer protocols are by definition regularized, calculated, and formal. However, it's worth reiterating, astute creators of user experience (UX) design and digital interaction know that not every aspect of a digital system needs to reside behind a firewall requiring the user to recall a password. Digital encryption operates at many levels in a system, most of which are invisible to the end user as data courses through networks. Some access is negotiated via passwords managed by the operating system, or via face recognition software and other biosensing technologies. The visibility of digital encryption protocols is as nuanced as people-centered urban design. The cryptographic city does not need to be the *regulated* city.

Crypto-Urbanism

The encryption of communications and data flows is now commonplace and goes largely unnoticed by regular consumers and users. But there was a time when cryptography subverted norms.

As this book is about the built environment it is hard to resist a brief foray into the architectural meanings of *norms* and *rules*. The word *normal* is derived from the Latin noun *norma* which was the square of wood used by carpenters and masons for creating right angles.[36] It appears among the symbols of Freemasonry. As known to any student of geometry, a line (or wall) is *normal* to another if it meets it at right angles. Common usage has since generalized *normal* to refer to any operation or behavior that follows a pattern, template, or model. *Rule* originated from the Latin *regula*, a rod for drawing straight lines and measurement, though it's adopted now to indicate prescriptive statements about behaviors that are to be followed. Rules can be recited or written down, regularized—as regulations. It is hardly surprising that spatial expressions of rule breaking often settle on sloping walls and angular shapes and geometries that are non-orthogonal.

Codes and *code breaking* come to us within a more legalistic frame.[37] Urban geographer Rob Kitchin means by *code* the instructions in computer programs that increasingly govern our lives: "code and space are mutually constituted,"[38] by which he means that code is "both a product of the world and a producer of the world."[39] Digital code pervades our urban experience. Urban theorist Stephen Graham makes similar observations by referencing "the hidden world of codes": "code-based technologized environments continuously and invisibly classify, standardize, and demarcate rights, privileges, inclusions, exclusions, and mobilities and normative social judgments across vast, distanciated, domains."[40]

A book by urban planning scholar Eran Ben-Joseph, *The Code of the City: Standards and the Hidden Language of Place Making* presents *codes* as "regulations" that "exert influence and shape the global landscape."[41] Building codes and standards, many of which have changed over time, are indeed hidden or latent as norms within the fabric of a place. According to common usage as I have outlined already, norms follow a pattern that may be invisible; rules are explicit.

The extremes of rule-breaking behavior or living without rules is captured well by the concept of *anarchy*—being without a ruler or rule.[42] Breaking into secret messages strikes anyone with a mind for rule and order as an act of defiance and disorder. But the movement of the 1980s known as *cypherpunk* championed cryptography as a means of evading government and state instruments that might seek to monitor its citizens.[43] Digital encryption methods would enable "individuals and groups to communicate and interact with each other in a totally anonymous manner."[44] The cryptographic methods formerly the preserve of the U.S. National Security Agency (NSA) and other state bodies would be distributed and improved for anyone to create secure messaging: "crypto anarchy will create a liquid market for any and all material which can be put into words and pictures."[45] I am quoting here from the "Crypto Anarchist Manifesto"[46] put forward in 1988 by the founder of the movement, Timothy May.

A 1993 article in *Wired* about the movement captured its mood, reporting the dream that an "individual's informational footprints" "can be traced only if the individual involved chooses to reveal them; a world where coherent messages shoot around the globe by network and microwave, but intruders and feds trying to pluck them out of the vapor find only gibberish; a world where the tools of prying are transformed into the instruments of privacy."[47] In these early analyses, cryptography was seen as an enabler of free, grassroots, trusted, and neighborly information flows. The cryptographic city is caught up in these disruptive practices, overtly in the case of political activism but also breaking in and theft.

Urban Disruptors

There are many ways of disrupting the orderly arrangements of a city.[48] One disruptive practice is to welcome outsiders who introduce "the shock of fresh habits and ideas: challenges to old ways" according to Lewis Mumford in *The Culture of Cities*.[49] Other disruptive practices seek to override the usual arrangement of things and impose new methods of organization, or of breaking through the organization. They turn access into a challenge about how to bypass and subvert the spatial arrangement of the city by creating alternative entry points. They try to break in.

Mumford described the ideal city as the *metropolis*, "the mother city."[50] In place of the *polis*, consider the term *borough*, a word of Germanic origin. It denotes a fortified "civic community" according to the *Oxford English Dictionary* (hereafter, OED). Now *borough* usually implies some kind of protection or fortification, such as a castle, a court, or a manor house—and the related proper nouns ending *burgh*, as in *Edinburgh*.

In native English a burgh-*breche* was a break-in, now contracted to *burglary*, "The crime of breaking (formerly by night) into a house with intent to commit felony" (OED). The term was usually applied to a house, a domicile, a residence, a place of habitation—and now that includes the "people's house," the U.S. Capitol. Burglary implies rule breaking and a form of anarchy, and is in the company of theft, arson, robbery, vandalism, insurrrection, and other crimes affecting person and property, including cybercrime.

These definitions bring burglary under the purview of the city. At least that is the proposition advanced in Geoff Manaugh's *A Burglar's Guide to the City*.[51] He develops the theme by recounting the history of George Leonidas Leslie (1842–1878), a Cincinnati-trained architect who turned to robbing banks, mostly in New York. As described by Manaugh, Leslie's approach to burglary was inventive, and even theatrical, demonstrating a clear understanding of place and space. He would create full-scale replicas of the places he and his gang intended to rob in order to leave little to chance, a familiar heist scenario: "Pirates of space-time, dressed in opera costumes, picking bank locks and assembling duplicate vaults in abandoned Brooklyn warehouses, Leslie's gang and their astonishing success rate set a delirious precedent for future burglaries to come. Leslie thus became both burglary's patron saint and architecture's fallen superhero, its in-house Lucifer of breaking and entering."[52]

Manaugh makes the case that concepts derived from the nefarious practices of burglary have informed our conceptions of the city, and even the form of our cities. Continuing the story of the Cincinnati architect-criminal: "You cannot tell the story of buildings without telling the story of the people who want to break into them: burglars are a necessary part of the tale, a deviant counternarrative as old as the built environment itself."[53]

To view places through the practices and eyes of a burglar might indeed provide new ways of looking at the city and its architectures, and has the potential to influence urban design, as architects and city planners install the means to prevent burglary but as they contemplate burglary's spatial and temporal implications. If crime is hardwired into the history of architecture, mainly through measures to prevent crime, then so is cryptography. My claim that cryptography as a means of concealing and revealing information is hardwired into the fabric of the city finds further justification.

Quiet Places

In chapter 2 I focused on text as written or printed, and alluded to deictic utterances, shibboleths, and quirks of pronunciation. There's much that could be said about *crypto-sonics*, sound as a coded medium. In order to close off the discussion I began in chapter 2 about writing and place I want to now conclude with whispers.

Whispering provides an obvious connection with secretive communications, speech directed intimately at a designated recipient, in resistance to Eve the eavesdropper. Steven Connor has written extensively on the cultures of the voice. He says in his book *Beyond Words: Sobs, Hums, Stutters and Other Vocalizations*: "The whisper signifies intimacy and secrecy. It is the mode in which I most naturally speak to or overhear myself. As such, it has religious or supernatural overtones, the whisper being the favoured mode of communication both of angels and of demons. The intimacy of the whisper gives it strong erotic force, too, as in the many popular songs in which whispering features."[54] The popularity of whisper videos[55] also demonstrates the long-standing fascination we humans have with the voice speaking softly.[56]

Whispers aren't as secretive as we might think. More than concealing a message, they announce that a secret is being told. Some urban spaces make it easier to hear whispers, such as beneath domes or in caves, and these are sometimes called "whispering galleries." Nooks, cosy firesides, and well-appointed corners of rooms and urban spaces encourage soft and intimate talk. Some spaces incline us to whisper, due to the solemnity or acoustic reverberation of a space. While researching this theme I happened to visit the town of Echternach in Luxembourg. The peaceful crypt of the abbey was suffused with the gentle slurping and gurgling from a stone channel fed from St. Willibrord's Spring. It sounded as a whispering voice. Whispers and whisper-like sounds seem to belong in such places and engender a kind of calm reverence.

Whispers also have a disconnected aspect. Connor associates whispering with ventriloquism: "The whisper is this voice, embodied, but without abode."[57] So whisper spaces might be un-homely, or detached spaces. More profound perhaps is the provocative assertion that to whisper is to play between the spaces of inside and outside.[58] Connor goes on to suggest that this ambiguity of interior versus exterior contributes to a whisper's purchase

in the realms of secrets and rumors, "the shades of the underworld."[59] In
this respect, whispers equate to dark, shadowy places,[60] further aspects at
the hidden reaches of the cryptographic city.

Subliminal Codes

Beneath the radar of even whispered speech are messages that evade conscious attention. Traditionally, cryptography operates with single messages.
It has the force of a command or an instruction, a singular performative message given added weight through its mode of delivery. Who could ignore an
encrypted message from high command! But communications also impress
their target through frequency, by constant repetition. A message can be
rendered invisible as background noise, as in an advertisement repeated ad
nauseam on television or at the start of a YouTube clip, but it is no less force-
ful. Like most effective propaganda, such messaging operates "subliminally."

The classic experiments in subliminal messaging were less about covert
signals hidden within innocuous public service announcements, and more
about plain text messages concealed within advertisements. Audiences in a
cinema could be induced to buy a Coca-Cola during intermission if presented
with a few images of the brand and texts telling you that you are thirsty.
These suggestions would appear for a split second in the film as just single,
isolated frames evading conscious detection. These techniques were not put
into widespread practice, but they were experiments by the marketing psychologist James M. Vicary in the 1950s.[61] The theory was that we pick up
words and images presented to us in ways that we neither notice nor recall.

There's no real evidence that this kind of subliminal messaging works
with the immediacy as proposed. By most accounts, the experiments were
a fabrication, though the idea of subliminal messaging persists as a cultural
meme, bolstered by ideas about mass hypnosis, charismatic leaders, irratio-
nal cults, and adherence to unlikely conspiracy theories.

Image technologies from moving film to digital processing encourage
suspicion of subliminal messaging. Thanks to video formats such as animated GIFs, images flashed in rapid succession are commonplace in the
digital age. If you have sufficient visual acuity you can race through your
own digital photo-library at about ten frames a second and identify the
portrait of your best friend. Images may also flash before your eyes of which
you have no recollection or awareness.

Though advertisers were skeptical about subliminal messaging techniques, the idea became a target for critics of mass media advertising. In his book *The Hidden Persuaders*, Vance Packard sought to expose the surreptitious nature of all advertising tactics: "These efforts take place beneath our level of awareness; so that the appeals which move us are often, in a sense, 'hidden.' The result is that many of us are being influenced and manipulated, far more than we realize, in the patterns of our everyday lives."[62]

The idea of subliminal messaging gained prominence after World War II as people reflected on the power of propaganda. A 1958 edition of *Life* magazine included an article featuring so-called "Hidden sell" techniques.[63] Film producers were on the verge of producing SP (subliminal perception) movies. A film might flash images of a skull at audiences to heighten the impact of a horror scene. The advertising scenario was best captured by a cartoon that showed a TV addict reclining in his armchair while his female companion looks on. Hair products are on the coffee table and his hair is in curlers. "I don't know what came over me," he says. The article highlighted serious and nonserious alarm about subliminal messaging.

What isn't subliminal? The idea that advertising might operate subliminally assumes that the application of such techniques is exceptional—as if we are otherwise in control of what we see, hear, and sense, and the effects those sensations have upon us.[64] Media presentations that reinforce stereotypes are also a form of subliminal messaging. It is a common feature of interpretation in any age that we are never fully aware of what influences our judgments and interpretations.

Such covert messaging operates by evading detection. It also operates through repetition. A single ad for wash-and-wear shirts may fail to have a direct impact but is more likely to if encountered repeatedly and across different media channels. With consumer profiling advertisers present readers with targeted messaging, repeatedly. I will continue the theme of repetition and combinatorial complexity in chapter 4.

In this chapter I've added a tighter focus to the claim that cities can be understood through the lens of cryptography. This took me on a journey through semiotics, theories of affordance, debates about urban legibility, norms, rules, notions of conflict, defiance, rule breaking, sonic practices, whispering, and hidden persuasion. Insofar as these themes speak to the affordances of cryptography, they bolster my case that the city is cryptographic.

II Urban Combinatorics

4 Urban Multiplicity

In his classic book *The Culture of Cities*, published in 1938, Lewis Mumford observed that cities are places where "the goods of civilization are multiplied and manifolded."[1] Cities benefit from mass concentrations of markets and the means of production. In so far as it involves physical spaces, the cryptographic city is a site of intricate combinations, permutations, and arrangements of a multiplicity of zones, buildings, rooms, streets, plazas, gardens, grids, signs, and other infrastructural elements. Combinations compound the challenges of multiplicity. The greater the number of books in my bookcase, the larger the number of ways they can be arranged: by size, color, author, title, and randomly. There are over three million alternative orderings of just ten books on a shelf.[2] Any spatial arrangement—including the organization of apartments in a multistory tower block or houses in a street—deals in combinations. Cities are more than just the arrangements of elements, but I would like to indulge this simplification for the purposes of this chapter, which is to establish further consonance between cryptography and the city.

Throughout this book I will refer to the *hash*, which is a prominent term in computing and cryptography. According to the OED, the *hash is* "so called because it consists of small pieces of code arranged in an apparently jumbled and fragmented way." I'll make use of its formal definitions in chapter 10, but it is relevant to concepts of multiplicity. To hash is to chop up, to hack, a term applied readily to food (recooked and chopped meat), narcotic dried herbs (hashish), and through further linguistic coincidence resonates with software hacking as applied drudgery and routine work, low-paid piecework, and labour delivered on demand in the gig economy.[3] To hash is to arrange, rearrange, or jumble elements in such a way that they are unrecognizable from the original arrangement. In architecture, art, and urbanism the concept of the hash most likely invokes concepts of *collage*, the deliberate or accidental arrangement of disparate elements to synthesize something new. According to architectural writer Jennifer Shields, "A collage as a work of art consists of the assembly of various fragments of materials, combined in such a way that the composition has a new meaning, not inherent in any of the individual fragments."[4]

Cryptography relies on such seemingly random arrangements. Anyone invested in securing property or money is familiar with the combinatorial complexity involved in opening a combination lock, getting through a security door, or releasing funds from a debit card account. The odds of gaining access by entering numbers randomly are diminishingly slim as the number of symbols in the code increases. Cities are protected in various ways by combinations.

As I'm pressing the case for the cryptographic city, I want to discuss the combinations of city elements and combinations in security codes, and to show how they each contribute to the arrangement of cities. Combinations not only are characteristics of access systems but also are endemic to the organization of the city.

In a chapter appropriately titled "Deciphering and the Exhaustion of Recombination" in her book on cultures of cryptography, Katherine Ellison states: "These material methods of transmitting intelligence illustrate the creativity of recombination and repurposing; books and bodies are manipulated so that their parts serve diverse functions; the familiar is disassembled and recombined to produce something new and only discernible to the senses trained to perceive it."[5] The novelty engendered by combinations and recombinations provides opportunities for hiding and disguising things, in books and in cabinets with secret compartments.[6]

If disassembling and recombining cabinets and texts serves to both hide things and to create something new, then so does the city. Add to this affordance the security aspects of recombination. Multiplying and compounding compartments and rooms provides a means of increasing the security of a place. As I'll show in chapter 11 on obfuscation and espionage, multiplication serves as a means of planting confusion, which is in turn a security measure. Confronted with an array of hundreds of hotel rooms or offices, an assailant would have to undertake some effort to find the target by searching every room.

Urban Combinatorics

Consider a basic security device, a physical combination lock. A combination lock is made up of a series of metal disks with grooves and notches in them. The disks move freely around a common axis but there is only one alignment of the disks that will permit free movement of a spindle in order to engage or release a catch. Key-operated pin-and-tumbler locks work on a similar principle of alignment. The combination of notches and grooves along the key have to correspond to the position of a row of spring-loaded pins of different lengths in the locking mechanism. When you slide your key into the lock you position it within a cylinder. The correct key in the lock allows the cylinder to rotate and engage or release the latch.

Elements rotating around an axis and locking into place are invisible in everyday lock and key operations, but it's a visual trope in sci-fi and fantasy. I'm thinking of *Indiana Jones and the Kingdom of the Crystal Skull* (dir. Steven Spielberg, 2008) and *The Mummy* (dir. Stephen Sommers, 1999) as the explorers finally put the last parts of the key together to cause a panel to rotate and reveal the secret treasure. The film *Army of Thieves* (dir. Matthias Schweighöfer, 2021) fixates on cylindrical lock-and-pin movements as the virtual camera flies through the mechanisms of a series of safes that are themed on Wagner's *Ring Cycle*. I think also of architecture: the Nakagin Capsule Tower by Kisho Kurokawa and the Shizuoka Press and Broadcasting Tower by Kenzo Tange and others of the Metabolist school.[7] Nothing moves, but the living pods are locked into place around a central service axis as if they could.

The numbers visible on the spindle of a combination lock constitute a key, as do the invisible pin lengths on a key-operated pin-and-tumbler lock. As evidence of some people's fascination with locks consider those who have turned picking locks into a hobby. TOOOL stands for The Open Organisation Of Lockpickers. They hold an annual event called LockCon. The organization distances itself from criminality, actual breaking and entering. It provides instructions and runs competitions as a kind of "door hardware sport," according to their website. But the main source of fascination resides with the idea of the lock as a puzzle to be solved: "Lockpickers see locks as puzzles, and solving such a puzzle provides an enormous thrill ;-) This thrill motivates people to carry on with it, and try an even more difficult lock. It is addictive, but pacifying all the same."[8] They also claim to expose vulnerabilities in manufactured locking systems. They thereby claim to provide a public service, also helping people who have legitimately lost their key to the front door or safe. The group admits, and seems to enjoy, that it is under suspicion, and they claim that agents of the law attend their events under cover. That the organization presents as a puzzle to others heightens the thrill.

Combinations are of interest as puzzles to architects and planners, and not just as they labor over key schedules.[9] Think of the arrangement of rooms in a floor plan, perhaps a hospital floor plan, with many functions, relationships, and constraints. Drawing an analogy with locks we might think that one arrangement provides the answer, the solution, the key to the planning problem. Floor plans are also a bit like jigsaw puzzles. But in a jigsaw, there's only one combination of spatial elements (jigsaw pieces) that reveals the (hidden) picture, the solution to the puzzle. In the case of floor plans there's not a single combination that affords a single best arrangement, a solution. Nor are all rules and criteria for the method of combination clear at the outset. We tend to think of a combination that provides an optimal condition, the best arrangement all things considered.[10] That's an operation in combinatorial complexity, though substantially less precise and well defined than arranging jigsaw pieces or aligning spring-loaded pins in a lock to allow a cylinder to rotate. Planners, designers, politicians, developers, citizens, and the forces of nature configure, organize and disorganize cities as combinations. By this formalist analogy, combinatorial complexity runs deep in the structure of the city.

To combine and enumerate are mainstays of puzzles, and of much that people think of as rationality. In chapter 3 I alluded to clues and evidence in the environment as triggers by which we recognize affordances, as part of the "code" of a place. In a forensic context those clues have to be pieced together, combined, recombined, and filtered. The writer Arthur Conan Doyle had Sherlock Holmes assert, "When you have eliminated the impossible, whatever remains, however improbable, must be the truth"[11] One of the ways to eliminate the impossible is to first enumerate everything, that is, all possible event sequences and motivations that can be enumerated—probable or not. The doyen of rationalism René Descartes said something similar. His last rule for sound reasoning "was to make such complete enumerations and such general reviews that I should be sure to have omitted nothing."[12]

It is as if the solution to a philosophical puzzle is to lay out all the possibilities and in all combinations in order to alight on the most clear-sighted. Cartesian rationality was one of the pillars of systems theory and those scholars of the Design Methods Movement who would seek such orderly, mathematical, and logical procedures for designing buildings and laying out cities.[13]

Lionel March was a pioneer in mathematical methods applied to architecture and design, among other notable contributions. His 1971 book with Philip Steadman, *The Geometry of Environment*, and subsequent books and articles makes clear the value of combination and enumeration, laying out all possibilities and permutations.[14] *The Geometry of Environment* incorporated the New Math of the 1960s and *set theory* into thinking about architecture and space. Their study was directed mainly at floor plan layouts: enumerating the possible ways that rooms can be arranged in a building taking account of adjacencies and connecting doorways. As a more abstract challenge, this could be all the ways that a rectangle can be divided into a series of tightly packed smaller rectangles. They called the process "rectangular dissection."[15] Once all possibilities are enumerated, it is then possible to count, classify, order, and filter such arrangements, and derive their properties. Such enumeration is also something you can program computers to do.

Then there's the challenge of permuting very large numbers of elements such as rectangular rooms. According to Bloch and Krishnamurti working on a similar project, there are over 280,000 ways that just ten rooms can be arranged bounded within a rectangular perimeter and ignoring room dimensions.[16] For greater numbers of rooms the number of possibilities becomes unwieldy to enumerate and process.

Under the influence of *The Geometry of Environment*, those of us with a computational bent were fascinated by the idea of enumerating all possible combinations of building elements, such as rooms in a floor plan, classifying them, filtering and selecting from all possible permutations. I had my own foray into rectangular dissections, and an attempt to deal with large numbers of rooms that meet some kinds of relational constraints as I outlined in *Logic Models of Design*[17] (figure 4.1).

By most accounts, in spite of Descartes' formulation, rational thought does not actually proceed by way of exhaustive enumeration and combination.[18] Considering the difficulties of such enumerations, and questions

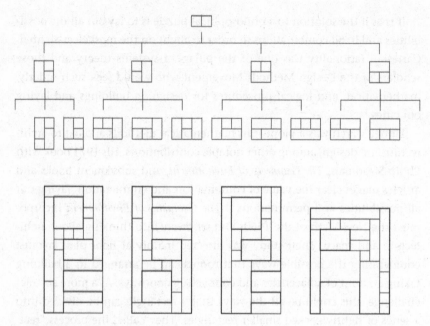

Figure 4.1
The start of a derivation tree of rectangular dissections resulting in spatial combinations of rooms and courtyards in plan. *Source:* Author.

about their practical usefulness, one may well ask what motivates such an interest in enumerating combinations of elements in architecture and urbanism. From a phenomenological and psychological point of view, I propose that apart from any practical use, such enumeration fulfills several human needs and desires relevant to my analysis of the cryptographic city.

A Phenomenology of Combinations

First, consider the need to collect and classify, which in turn demonstrates a desire to have mastery over a domain of expertise. If you can enumerate, then you can control. I mentioned the research of Walter Ong in chapter 2. In his account of the European trajectory toward organization and the pretense of rationality we see double-entry bookkeeping, but also the classification of information and knowledge, and the emergence of the encyclopedia as a teaching tool.[19] Such cultural innovations were in the company of botanical and animal classification, enumerations of architectural styles and urban typologies, and other manifestations of encyclopedism. Laying things out in order is a way of exercising control.

Second, we are drawn by the allure of rhythm: seeing patterns, repeated patterns, and repetition in patterns.[20] Enumeration of combinations sets up a rhythm as in a production line, a poem, music, or a riddle.

Third, we're motivated by the anxiety that we'll miss something essential. As Descartes said, he thought it necessary in his philosophical reflections to provide such complete enumerations to "be sure to have omitted nothing."[21] Descartes couched his *Discourse on Method* in terms of self-referential anxiety. Fear of missing out is a basic human trait. It is the elusive combination that drives the restless generation of yet more permutations, and the search for answers and "solutions."

Fourth, we may be afraid we'll fail to find the crucial combination, the key to the safe as it were. Combinations provide a key to unlock something hidden, as in the case of a puzzle. Combinations provide a way of concealing and revealing mysteries, and of making the ordinary mysterious. The impulse to enumerate participates in the mindset of the habitual gambler: how many combinations of cards can there be before I hit on a royal flush? There's a moment in the Korean horror series *Squid Game* (dir. Hwang Dong-hyuk, 2021) where the characters have to cross a bridge made of adjacent pairs of glass panels. Only one of each pair of panels is safe to stand on. At one moment in the perilous crossing one of the characters positioned partway along the bridge realizes the extremely slim odds he has of getting to the end. He has fifteen pairs of panels in front of him. Terrified, he realizes his chance of survival is "two to the power of fifteen. . . . That's a 1-in-32,768 chance!"[22] Sometimes finding the right combination against the odds is a matter of survival and ignorance of the odds provides false comfort in the face of the inevitable.

Combinations and Riddles

Riddles operate via combination and permutations. The Sphinx was a mythic trickster, a riddler, that guarded the gate to the city of Thebes and required travelers to answer a trick question before gaining access to the city. Here, the riddle serves as a passcode. Riddles typically present as permutations, a combination of elements, albeit for small numbers, usually around 2, 3, or 4, or at least a subset of possible combinations. The Riddle of the

Sphinx asked, "What is the creature that walks on four legs in the morning, two legs at noon and three in the evening?" The riddle draws on permutations of morning, noon, and evening and the number sequence 2, 3, and 4. By most accounts, the key to resolving the riddle is to appreciate that morning, noon, and evening could apply to the early, middle, and final stages of life, and that "legs" could apply to arms and walking sticks as well. So, the resolution to the riddle of the Sphinx presented to the traveler is "a man."[23]

Urban combinatorics, or assembly, is not only a matter of deriving the best solution as if solving a puzzle. Permutations increase the chances of encountering an incongruity. After all, it was the tactic of the Surrealists to rearrange and juxtapose familiar elements in unfamiliar (incongruous) ways, as collage. They felt no embarrassment in combining sand dunes, obelisks, and crucifixes with elephants on stilts.[24] The permuted riddle format of the Sphinx also fulfills the criterion of incongruity. The permutations selected for the Sphinx riddle work as an exercise in incongruity, if not absurdity.

A senior citizen with a walking stick is less engaging as a resolution to a riddle in the current age. Poking umbrellas up chimneys has greater potential: What goes up a chimney down but cannot go down a chimney up? That's also a riddle that begins with permutations—of "up" and "down." To the child it's the combinatorial, repetitive, and rhythmical aspect of the riddle space that provides the initial appeal—even before the child appreciates the mechanics of the circumstances, and the ambiguity and its resolution that provide the sense of the absurd, and a joke.

I'm here considering riddles because they involve permutations and combinations. It is satisfying to think that with all its mathematical complexity, contemporary cryptography begins with the riddle, or at least a myth (of the Sphinx) involving a riddle as a rudimentary combinatorial challenge. Riddles are a kind of puzzle or key. After all, the Sphinx required an answer to the riddle before the traveler could enter the city. To enter the city, you had to answer the riddle correctly—or die trying. The challenge was not to establish a correct combination, but to provide a key, the answer, that ensured that the combination made sense, and resolved or enhanced the incongruity.

By one theory, the transition from the incongruity in the statement of the riddle to a resolution constitutes an aha moment. The transition from confusion to clarity provides a moment of enlightenment, satisfaction, pleasure, and even delight.[25] The riddle also presents something extraordinary, impossible, or monstrous, like the Sphinx itself—a lion with the face of a human being—and transitions to something reassuringly ordinary.

To reiterate the role of combinatorial intricacy in the urban context, buildings and cities involve combinations of elements, spaces, rooms, furniture, and functions. In this respect they form the basis of a riddle, or at least a puzzle. A floor plan follows the format of a kind of puzzle, as a designer creates it, and as visitors move through it. The experience of the city also presents as a riddle, at least for the first-time visitor. Movement through a city can have such a character—moments of confusion followed by moments of clarity. Confusing, disorienting or jumbled spaces transition to open, clear, ordered places. Permutations of elements in the visual field that present contradictions transition to an overview that shows that the whole place makes sense after all. Think of moving through the tangle of streets in an old medieval city, and the need felt by the fit and able to climb to the top of a tower to see how the city looks from above, to make sense of the jumble of relationships experienced at ground level and discover that the cathedral is around the corner from the town hall, which is adjacent to the cafe you just visited. There's pleasure in that, the contrast and the transition—like solving a riddle.

Hacking the Combination

I have mentioned the Bletchley Park codebreaking project a few times. It provides a potent illustration of the combinatorial challenge that has cryptography at its focus. The UK's World War II codebreaking headquarters at Bletchley Park opened to the public in 1994. Those of us with an interest in computing went there to pay homage and to see the ranks of humble low-rise buildings and examine firsthand some of the codebreaking devices, to learn and to imbibe the enthusiasm for cryptography and cryptanalysis of the volunteers who populated this living museum.

The Bletchley Park museum teaches us that during World War II German operatives at command stations would be handed messages to be encrypted before relaying them via Morse code to U-boat commanders. The operative would enter each character of the plain text message into a tabletop machine—the commercially produced Enigma machine—that looks something like a typewriter. With each key stroke, a different letter from the one pressed would illuminate on a panel at the top of the machine. An operative would have to write down the new characters as they appeared. The message so encrypted would then be sent in Morse code via conventional telegraph to the recipient, who would then decrypt the message by means of a similar machine, and a similar process. It didn't matter to the sender and recipient that the telegraph service could be tapped, as the interceptor would only pick up an apparently random sequence of characters.

The Enigma machine didn't map every character in a unique and predictable way, as if every "T" got encrypted as a "P" as in a substitution cipher. That would be trivial to decrypt. Each character was translated through several scrambling operations involving a series of disks that rotated with each press of the keyboard. The machine operation was a version of Alberti's cipher disk mechanism, though the disks would move relative to each other with each letter. The message recipient could decrypt the message as long as they had the same Enigma machine, with the same disks in the same positions, the same starting conditions, and the same circuit. The sender and receiver also had the same code book, which listed the required parameters for the machine on any particular day. Any interceptor with a similar Enigma machine and the code for the day could decrypt any messages they picked up.

The codebreaker challenge for the Allies as cryptanalysts was to develop a system for deriving the code for the day. This required cunning, experimentation, and electromechanical devices that could iterate through very large combinations of parameter configurations. The first such machine was known as "the Bombe," a proto computer. It was crucial that the Germans did not know how their messages were intercepted. A few changes in the encryption method would have meant the codebreakers would have to revise their methods to uncover that. As I outlined in chapter 3, as well as the contest of a devastating war, cryptography itself was a site of contest between coders and codebreakers. It was also a contest against combinatorial complexity. Bletchley Park is a reminder of how the securing of secret information relies on the sheer quantity of possible combinations of numbers, symbols, and sequences.

Counting Characters

Combinatorial complexity is one of the means of confounding codebreakers. But consider how codebreakers have exploited the combinatorial aspects of language, understood through letter frequencies. As I have shown, a substitution cipher is one of the simplest methods for encrypting a message. A unique symbol stands in place of each letter in the hidden plain text message. The symbol set can consist of any arbitrary set of characters, as long as each symbol maps uniquely to the letters of whatever alphabet you are using for the plain text message. The usual encryption method is to deploy a different letter from the same alphabet, so that the encrypted message contains the same character set as the hidden message (e.g., the twenty-six letters of the English alphabet A to Z plus a space). This method is often referred to as the Caesar cipher. It is also the method deployed in the Masonic (pigpen) cipher and Wittgenstein's cipher.

As known to any Scrabble player, letters in a block of text occur at different frequencies. For example, E is more common than Q. Letter frequency provides a clue to decrypting a coded message. If the frequency of the letter E is 12.1 percent, then the probability that the letter E will show up in any position in a coded string is 0.121. On the *Practical Cryptography* website, James Lyons provides a helpful blog post with the frequencies of letters and in various combinations.[26] As text messages are sequences of characters, it would also be useful to know the probability that any letter will be followed by any other letter. How often is an E followed by another E, or an S or D, and any other letter? According to Lyon's frequency data, in any block of text you can expect the letters EE to occur in 3.54 percent of all adjacent pairs of letters, ES occurs 1.32 percent of the time and ED 1.08 percent of the time. I calculated these percentages from Lyons's letter frequency data.

A cryptanalyst might also want to know the frequency with which, given an E, the next letter will also be an E, an S or a D, and so on. With that statistical data we can produce probabilities that populate a transition matrix, that is, a table showing all the letters of the alphabet plus the space character, and the probabilities that any letter will be followed by any other, including itself. That information can also be understood as a transition network with twenty-six nodes and a tangle of connecting arrows connecting each node—and with probabilities attached. I show a more manageable subset of the challenge in figure 4.2.

This information about the probability that one particular letter will be followed by another particular letter provides the ingredients for a Hidden Markov Model (HMM) formulation of the problem of automatically deciphering a substitution cipher. In HMM terms, the hidden part is the path

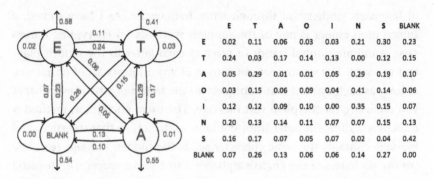

	E	T	A	O	I	N	S	BLANK
E	0.02	0.11	0.06	0.03	0.03	0.21	0.30	0.23
T	0.24	0.03	0.17	0.14	0.13	0.00	0.12	0.15
A	0.05	0.29	0.01	0.01	0.05	0.29	0.19	0.10
O	0.03	0.15	0.06	0.09	0.04	0.41	0.14	0.06
I	0.12	0.12	0.09	0.10	0.00	0.35	0.15	0.07
N	0.20	0.13	0.14	0.11	0.07	0.07	0.15	0.13
S	0.16	0.17	0.07	0.05	0.07	0.02	0.04	0.42
BLANK	0.07	0.26	0.13	0.06	0.06	0.14	0.27	0.00

Figure 4.2

Transition network and table. The network on the left shows the three most frequently occurring letters (and a blank space) in the English language, and the probabilities that one letter will follow another in a word. The table shows the same information for the seven most common letters. The probabilities here are derived from anagrams of ETAOINS generated at unscramblex.com. The matrix and network would be much larger for all the letters of the alphabet. Such sequence representations are useful in codebreaking, DNA sequencing, weather prediction, machine learning, and other applications of Hidden Markov Models (HMM). *Source:* Author.

through a network of twenty-six letters (plus space) that make up the order of letters in the hidden message. The observation part of the HMM formulation is the encrypted version of the message. I'll also say more about Markov models in chapter 12. Here, it is sufficient to note that information about the frequencies of letters, and the frequency of particular combinations and sequences of letters forms part of a typical codebreaking toolkit.

The variation in letter frequency was important initially in movable type printing. You needed more Es than Qs, so a printer's "type case" would need a bigger compartment to hold the letters. Figure 4.3 shows an image of a type case that spatializes letter frequency as something that looks almost like a city block or item of furniture.

Calculating the letter frequency in a block of text provides a way of making a good guess at the language it's written in—automatically. An interesting website at letterfrequency.org provides the frequency order of letters in a range of languages.[27] For example, the frequency order in English starts "ETAOIN . . ." In German it starts "ENISRAT . . ." In French it's "ESAITN . . ."

In the 1950s, the pioneering information scientist Herbert Ohlman ran calculations on sets of texts to determine the relative frequencies of letters.[28] He calculated the frequencies in different parts of words: the first,

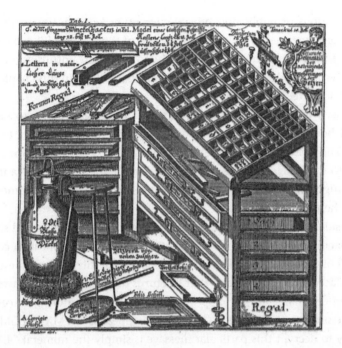

Figure 4.3
An eighteenth-century "type case" that spatializes letter frequency. The drawing is
captioned: "By Christian Friedrich Gessner—Illustration taken from a scan of: Christian Friedrich Gessner, 'Die so nöthig als nützliche Buchdruckerkunst und Schrift-
giesserey: mit ihren Schriften, Formaten und allen dazu gehörigen Instrumenten
abgebildet auch klärlich beschrieben, und nebst einer kurzgefassten Erzählung vom
Ursprung und Fortgang der Buchdruckerkunst'" (1740), 226f, and is in the public
domain via https://commons.wikimedia.org/w/index.php?curid=13538088.

second, third, fourth, and fifth positions. If we need any further support
for the importance of cryptography, Ohlman also asserted: "Coding, or the
transforming of information from one guise to another, is one of man's
commonest activities. Every picture may be said to be a coding of some real
scene and every written word a coding of some utterance—the brain itself
is said to work with coded impulses."[29]

Encryption Keys

Cryptography reduces combinatorial complexity for the sender and
receiver by the use of keys. For my purpose here it is sufficient to note that

the operations of public and private encryption keys continue this story about combinatorial complexity. In the rest of this chapter I'll explore the basics of their operation for any reader who shares my enthusiasm for the instruments of codes and permutations. So what follows in this chapter will involve some technical detail.

It is easy to grasp the idea that you need a key to gain access to a safety deposit box or a room to access something that someone wants to keep secret. The answer to the Riddle of the Sphinx is a key, which in turn gives access to the bridge and the city beyond. Cryptographic keys are a means of turning a seemingly arbitrary combination of characters into something comprehensible in plain text.

Consider a simple substitution cipher. The message is encrypted simply by replacing each letter in the text with the letter that appears a certain number of characters further on in the alphabet. For this purpose, the alphabet is placed around a circle so that it returns on itself. If the displacement is 4 then the characters ABC appear encrypted as EFG. Ignoring spaces, the message such as "WRITE ME A CITY" would appear as "AVMXIQIEGMXC." The key to decrypt this particular message is simply the numeral "4." The recipient just has to know that the key is the number 4 to apply the simple algorithm of counting back that number of letters. That is not very secure, however, and is easy for codebreakers to intercept without the key. They just need to keep adding or subtracting integers to work their way through positions in the alphabet until they stumble across a sequence of letters that looks like a coherent string of plain text. As well as letter frequency, codebreakers can exploit information about letters less likely to appear in pairs, such as A, H, I, J, K, Q, U, W, and Y.

Instead of a single number for the displacement key the encryptor could specify a series of numbers repeated across the message. A ten-digit key would increase substantially the degree of difficult for a codebreaker. So, a key 1437823605 means count up the alphabet 1 for the first character, 4 for the second, 3 for the third, and so on. That's like a rotating combination key on a padlock. The key then repeats for the rest of the coded message to turn my original plain text message into "XVLAMOAGCNUC." That would be impossible to decipher in a reasonable time using paper and pencil, though not for a computer iterating through every possible combination of numbers for a ten-digit key.[30]

Alphanumeric data is stored and transmitted in binary, as series of 1s and 0s. So, the counting process for the substitution cipher is replaced by flipping binary bits according to the sequence in a binary key. In this case the key is a series of 1s and 0s. Most digital encryption now uses 256-bit encryption keys. Cybersecurity expert Jon Watson provides a helpful explanation of the process that involves multiple iterations to effectively scramble the 1s and 0s, though in a way that allows the receiver to use the original sequence to restore the message with the same key.[31]

What I have described here is known as symmetric key encryption. The sender and the receiver of the message (i.e., the sender's and receiver's computers) use the same key. The AES (advanced encryption standard) is developed around this method. One of the main vulnerabilities of this approach is that the encryption key must be shared between the communicating computers before they can generate and transmit their encrypted messages to one another. The encryption key will need to be communicated via an insecure, or less secure, connection. If someone intercepts that transmission, then they will have the key and can decrypt the secret messages. It also means that the sender and receiver are both custodians of the key, and each will have a different stake in its security. So it is vulnerable. That's like giving a copy of your house key to someone to collect your mail while you are away. As the owner of the house I am likely to be more vigilant about protecting the key than the mail collecting friend would be.

Public Key Encryption

Public key encryption, also known as asymmetric key encryption, offers a solution to the problem of transmitting cryptographic keys across unsecured networks, the so-called "key exchange problem." In this method, the key to encrypt a message is different from the key used for decrypting it. It doesn't matter if other people intercept or even use the same encryption key as you do because that key cannot be used to reverse the encryption process.

The complicated chaining and nesting of access privileges in securing the exchange of data has spatial correlates. Various metaphors come to mind for public key encryption. It is as if you and your friends have identical keys that open the letter slot in your front doors. That is the public key.

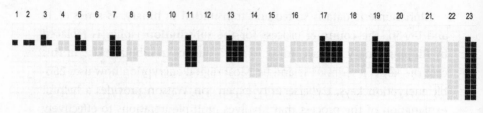

Figure 4.4
Spatial representation of the first ten prime numbers (in black). Public key encryption uses very large primes. *Source:* Author.

Any of these key holders can open the letter slot of any of the houses and drop a parcel into the house, but only the owner of the house has a key that gives one person access to the house, the owner. That is the private key. Once inside they can access the parcel.

How is this kind of security implemented in the case of digital data? Large numbers secure data: not just any large numbers, but primes. Digital public key encryption systems commonly deploy algorithms that exploit the properties of prime numbers to encrypt a message. A prime number is a positive integer that is not the product of two other integers (except 1 or itself). If you try and draw a rectangle on a regular grid that only takes up a prime number of grid units, there will always be at least a grid cell left over. Prime numbers are always odd (apart from the number 2). You can always turn a prime into a non-prime (i.e., a composite) by subtracting 1 from it (except for the first three primes in the series). As I'm keen to keep this account spatially relevant, I show the first ten primes as grids of squares in black in figure 4.4.

It is easy to multiply two numbers, even if very large, but more difficult to factor a number, meaning to find two numbers that when multiplied result in that number. It is especially difficult if the two numbers multiplied are primes. They will be the unique factors of their product. There is only one solution to the factor challenge for a number so produced. Factoring is a trivial calculation for small prime products, for example, the prime factors of 35 are 7 and 5 and only those two numbers. But factoring is computationally taxing for prime products over one hundred digits long.

Public key encryption requires two numbers: a public key number and a private key number. A helpful blog post by Nick Sullivan explains that "you can take a number, multiply it by itself a number of times to get a random-looking number, then multiply that number by itself a secret number of times to get back to the original number."[32] By "large numbers" cryptographers mean numbers in the order of four hundred digits. A helpful tutorial paper by Kathryn Mann amplifies further the scale of the combinatorial challenge: "The lifetime of the universe is approximately 10^{18} seconds—an eighteen-digit number. Assuming a computer could test one million factorizations per second, in the lifetime of the universe it could check 10^{24} possibilities. But for a four-hundred-digit product, there are 10^{200} possibilities. This means the computer would have to run for 10^{176} times the life of the universe to factor the large number."[33]

Here is some detail about the process. Person B (Bob) wants to send person A (Alice) a secret message and therefore asks A for A's public encryption key. Person A transmits A's public encryption key to B to facilitate this. Person B uses that public key information to encode a message according to the algorithm described as follows. Person A has the private key and can decode the message. Three items are transferred in sequence.[34]

1. B's initial low-security request for A's public key.

2. The public key transferred from A to B, which is low risk. It doesn't matter if hacker C (Eve) intercepts that transfer and finds out what the public key is. All that would enable C to do at best is to encrypt a message, not to decrypt a message.

3. The encrypted message from B to A, which is secure. A can read this with the private encryption key. The message could only be read by C if C had the private key. The private key is never transmitted and stays with A.

The algorithm behind this method of public key encryption is called RSA encryption after the three inventors, Ron Rivest, Adi Shamir, and Leonard Adleman.[35] These operations are of course invisible to the computer user and are the kinds of operations carried out by browser and server software in requesting and transmitting information securely across the Internet. The keys are also invisible to the users.

If it is not obvious by now, the encryption key is different from the short passcode that the user needs to access their computer, online banking, or account on a website. Passcodes are themselves encrypted before they get passed though the network to password servers that contain password lookup tables, which are in turn in a kind of code (a hash).

Combinations and Trapdoors

Public key encryption is like a turnstile or trapdoor, a one-way portal. You can go through it easily in one direction, but it is difficult to come out again in the reverse direction. A horizontal trapdoor relies on gravity for the one-way function. It is easy to fall through it into the cellar below, but harder to jump or climb your way out. That is similar to the way hunters use pits to trap prey, and some insects lure prey into a one-way system from which there's no escape. Trapdoors associate with secrets and hiddenness. A stage magician will install a secret trapdoor under a magic box. The assistant falls through the secret door before the magician opens the lid to reveal the box is now empty.

Cryptographers use the trapdoor metaphor to describe their methods. It is easy to encrypt a message, but has to be substantially more difficult to recover it without a special key, or escape route, as in the case of factoring a number into primes. It is easy to get the toothpaste out of the tube, but more difficult to get it back in. The main lessons from this mathematical exposition are that digital encryption takes advantage of combinations of number sequences, the irreversibility of some cunning mathematical operations, and the impossibly long odds of finding the right combination without clues, keys, and knowing the procedures and information about some essential parameters.

It's worth concluding this discussion with a further means of securing the process and some elegant mathematics. The multiplication of prime numbers can be made even more difficult to unmultiply if you redefine multiplication. One method is to define multiplication in terms of geometrical relationships between points on a curved line. Elliptic curves belong to a family of curves that make up the alluring surfaces of much contemporary organically inspired architecture. They are also the basis of encryption methods that secure digital communications. Mark Hughes provides a very helpful explanation of the Elliptic Curve Diffie–Hellman key exchange[36] that operates with primes and modular arithmetic. I provide an explanation that brings out some of the elegance of the curve geometries in some of my own online explorations.[37] I provide one of the diagrams in figure 4.5. If your computer and my computer are to communicate in secret then they need to agree to use the same elliptic curve, with the same parameters.

A great deal of modern cryptography is based upon the Diffie–Hellman key exchange, which requires that two parties combine their messages with

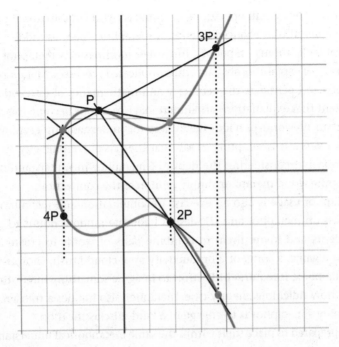

Figure 4.5
Elliptic curve with the general formula $y^2 = x^3 - ax + b$. It is possible to draw a straight line through two points on this curve that will only strike one other point on the same curve. Factoring into prime numbers is made even more difficult when multiplication is defined in terms of intersecting vectors on this curve. *Source:* Author.

a shared secret that is difficult for a bad actor to deduce. The Elliptic Curve Diffie–Hellman key exchange method allows microprocessors to securely determine a shared secret key while making it extremely difficult for an eavesdropper (Eve) to discover that same shared key. The method uses modular arithmetic commonly described as numbers that "wrap around" as on a clock face. A clock face has a "modulus" of 12. The prime number multipliers our computers choose for our private keys will also be many digits long as will the modulus and the coordinates of the initial point on the curve for the sequence of calculations, further securing the method.

Misplacing Words

Of necessity this has been a brief overview of important encryption methods. My main aim was to focus on combinations and how they can secure data flows. They entail operations in common with combinatorial puzzles. This chapter also introduced a series of metaphors: puzzles, riddles, locks, keys, cabinets, rooms, trapdoors, turnstiles, toothpaste. A metaphor serves as a means of explaining abstract and complicated processes. Metaphors also run deep and permeate the geometry and mathematics of encryption—to the extent that explanations and functional operations become the same.[38] The urban lifeworld provides metaphors for understanding cryptography and vice versa. It is tempting to say that the "cryptographic city" is just a metaphor, but considering the depth of metaphor in understanding and making the world metaphor simply reinforces the connection.

Metaphor usage is also an exercise in combination and recombination. In a class on metaphor and UX design I allocate random nouns to groups of students and invite them to combine pairs of words to create a concept for a game, a song, or a productivity app: cloud-brush, shower-hinge, umbrella-book. It requires little effort to imagine something interesting, no matter how ridiculous. In any case, metaphor also implies a transgression. To employ a metaphor is to engage in a kind misclassification.[39]

I'm prepared to place theft within the same metaphorical understanding. Cryptography aims to secure objects, places, and data against theft. To steal data is to recategorize what's yours as mine. Were it not for the despair and inconvenience we could see data theft as a profound exercise in the metaphorical imagination—or at least we can learn from it as such.

In *A Burglar's Guide to the City* Manaugh writes about a "hidden topological dimension tucked away inside the city."[40] Thinking of the clever bank heist that involves traversing rooftops, negotiating tunnels, and boring holes in walls, he suggests that "point A might illicitly be connected to point B"[41]; the burglar has to "make this link real" by means of "shortcuts, splices, and wormholes."[42] Criminal conjuring gives the illusion of spatial and temporal paradox: "Burglary reveals that every building, all along, has actually been a puzzle . . . a kind of intellectual game that surrounds us at all times and that any one of us can play."[43] This proposition that place is a puzzle adds further support for my assertions that the cryptographic city is a site of combinatorial practices, arranging and rearranging. Translating the theft of physical objects and its prevention to digital data further supports the proposition. Puzzles, hidden connections, and underground tunnels remind me of the workings of the labyrinth, the subject of the next chapter.

5　A Thousand Insides

Aspects of the cryptographic city are hidden from view, with wild and untamed depths. The cyber wilderness is also dark and deep. "Welcome to the Dark Net, a wilderness where wars are fought and hackers roam," wrote security advocate William Langewiesche.[1] Notably, this cyber wilderness is of uncertain extent: "The deep web is deep because it cannot be accessed through ordinary search engines. Its size is uncertain, but it is believed to be larger than the surface net above it."[2] Digital portals such as Tor (The Onion Router; see torproject.org) allow you to "browse privately" and "explore freely" via multilayered encryption and "thousands of volunteer-run servers known as Tor relays."[3] The Tor Browser promises that no one can track your online activity or identify your location as you access online resources, including via anonymous peer-to-peer (P2P) networks. Computer users can transmit large files P2P, including pirated software and media, and do so undetected—hence the underground black market connotations, especially if cryptocurrencies are involved.

The dark web is also a site of imagining particularly for those who have not entered its portals. Inventive minds fill voids and uncharted spaces. In his book *Imaginary Cities* Darran Anderson observes: "When faced with the blank space on the map, we turn to the fantastical."[4] He's referring to the maps of early explorers. Imaginative storytellers filled in those parts beyond the reach of their surveys with peculiar inhabitants, ghosts, monsters, and unlikely riches. To early imaginaries, cities that were off the map were places "where the inhabitants are perpetually drunk, where men eat birds and ride around on stags, where marriages are arranged between ghosts, and the lord in his marble palace drinks wine from levitating goblets."[5]

The subterranean parts of cities have similar blank spaces filled in by speculative invention. That explains in part our curiosity about life

Figure 5.1
LiDAR scan of basement book stacks and services at the National Library of Scotland. *Source:* Asad Khan, https://www.theentropyproject.com.

underground, as identified by Paul Dobraszczyk, Carlos López Galviz, and Bradley Garrett in their book *Global Undergrounds*. Having worked the surface, urban scholars turn to mining the depths, invoking terms such as a *vertical turn* and "the politics of subterranea."[6] It is easy to associate such spaces with crime. After all, that's where you can move about undetected from the surface. Geoff Manaugh observes this in the case of burglary: "The world is riddled with shortcuts and secret passages—we just have to find them. It is a crime, but it also symbolizes that there are ways of navigating the world that we ourselves have yet to discover."[7] Such underground haunts can in fact be prosaic places. Old and new cities have underground garages, cellars, tunnels, storage (figure 5.1) passageways, services, and communication systems, many of which are unused and obsolete.

I live in a street with a fifteen-meter-deep tunnel that for twenty-one years had a rail and cable system for hauling goods and passengers along its 1:27 gradient. When the cable system became obsolete, the tunnel was repurposed as an air raid shelter, and later as a mushroom farm. It is now sealed off. What's down below is difficult to see by peering through the substantial iron grille at its lower end. It is a hazardous place, and local authorities restrict access. The invisibility of underground facilities contributes to people's fascination with such hidden places. This particular tunnel features in a novel by Alexander McCall Smith who expands the underground network to nonexistent side tunnels and access points to coal cellars and buildings across the city.[8]

Paulsay Dobraszczyk and colleagues write about the underground parts of the English city of Nottingham in similar fantastical terms: "Nottingham sits above a burrower's fantasy: labyrinths within labyrinths of private artificial caves carved from the region's sandstone. Stairways in the backs of pubs, old cellars below houses, the lower reaches of shopping centres and car parks—all butt up against this hidden world."[9] Even if it weren't true, according to Anderson, "all great imaginary cities merge the matter-of-fact with the surreal."[10] We would want to believe in the city's underground passageways whatever the reality.

In chapter 1 I aligned the architectural crypt with cryptography as a major touchpoint in the cryptographic city. Novelists and filmmakers invoke the underground to explore the unknown. I recall episode 3, season 8 of *Game of Thrones* called "The Long Night" (dir. Miguel Sapochnik, 2019). Those who couldn't fight or were too important to lose in battle were told to hide in the crypt of the capital for safety. Meanwhile those above ground battled the indomitable White Walkers and their army of the dead. You shouldn't hide in a crypt when the Night King, the leader of the White Walkers, marches on the city. As the Night King slowly raised his arms in the midst of battle those who just died in the battle—or those long dead in a crypt—rose up and continued the fight as his relentless zombie army.

People may fear what resides underground, but there's a primordial connection there. In his fictional history of Middle Eastern oil extraction, philosopher and novelist Reza Negarestani invites the reader to fear and revere what's underground: "Ungrounded and unreported histories of the Earth are full of passages, vents and soft tunnels mobilized and unlocked through participations with the Earth as a compositional entity. These histories are engineered by openings and that which crawls within them; every movement in these passages invigorates the ungrounding of the earth, engineering what makes Earth, Earth."[11] Such autochthonous references connect the underground with earth, origins, and life.

Crawling, oozing materials and creatures and protohuman troglodytes live underground. They become disgusting when they move above ground as if matter out of place.[12] In a collection of short stories gathered under the title *Labyrinths*, the novelist Jorge Luis Borges (1899–1986) described

such an encounter: "In the sand there were shallow pits; from these mis-
erable holes (and from the niches) naked, gray-skinned, scraggly bearded
men emerged. I thought I recognized them: they belonged to the bestial
breed of the troglodytes, who infest the shores of the Arabian Gulf and the
caverns of Ethiopia; I was not amazed that they could not speak and that
they devoured serpents."[13]

Urban Labyrinth

Such underground spaces are not just voids and caverns, but also passageways with indeterminate connectivity. The labyrinth provides an
appropriate architectural type with which to characterize underground
places—invoking branched and looping passageways, dead-ends, losing
your way, navigating hazards and with no external clues as to where you
are or the scope and scale of the network. In chapter 1 I mentioned the
labyrinth as a motif for secret societies, and a motif of cryptography.

As I will elaborate later in this chapter, the format of the labyrinth
correlates with procedures for solving puzzles, including in video games
and some of the computerized processes in the cryptographic city and in
codebreaking. The philosopher and fiction writer Umberto Eco thinks that
detective stories are like a labyrinth.[14] I would add, so is the challenge of
trying out combinations on a lock, hacking, codebreaking, breaking and
entering, as well as instituting and defying security systems. A labyrinth is
also a means of organizing and reorganizing space.

Retractable queue barriers funnel tightly packed airport passengers in
twisted but orderly lines. These security labyrinths manage large numbers
of people within a confined space. They also keep people on the move, and
unceremoniously herd compliant travelers into conformity. Under social
distancing measures during the pandemic, these processionals became even
more common as people queued to enter airport departure areas, museums,
and supermarkets.

Umberto Eco identified three types of labyrinths. The most complicated
type is *rhizomic*.[15] You get caught in loops. Eco refers to sets of complicated
interconnecting passageways "where every path can be connected with
every other one. It has no center, no periphery, no exit, because it is potentially infinite."[16] I think Eco means that the rhizomic labyrinth is the world
as lived. You could think of a traditional market, a *souq*, in that way—the

common trope of the bazaar as organic, changing, multilevel, unplanned.[17] A tourist gets lost in the souq (and even wants to) to savor its sights, sounds, smells, and confusions—and to experience a kind of ludic vertigo.

Navigating the rhizomic labyrinth's twists and turns requires detective work. As a writer of mystery fiction, Eco gravitated toward the detective story as a labyrinth. The world in which the detective operates "can be structured but is never structured definitively."[18] Who among us is not at some stage a detective: sifting evidence, interpreting, and exercising indirect, abstruse, and abductive inference within a confusing landscape.

Eco's second type of labyrinth has junctions along the route where you have to make choices. There are dead-ends, and you may have to backtrack. Navigating such a labyrinth requires trial and error. Sometimes the routes through cities and shopping plazas are like that, especially for the tourist or infrequent visitor.

For Eco, the third and simplest labyrinth is the kind that has a center, and a single, convoluted path leads eventually to it. The traveler may be confused and lose orientation, but you can't get lost there. A security maze at an airport is like that. Technically, it is a *unicursal* maze. The traditional unicursal maze has no forking paths but leads inevitably through a series of turns to its center.

Labyrinths exist underground but also as drawings or patterns in tiling. As a drawing or ornamental pattern, you imagine the lines are walls and trace your finger along the path. You eventually arrive at the center through a series of left and right turns. There's just one path through the intestines of this maze structure. An illustrator designing a maze for a child's puzzle book will start by drawing out a single path with twists and turns from the start to the finish—center or goal—perhaps laid over an organizing grid. Once that route is established then it's only necessary to add alternative paths, dead-ends, and looping paths to make it more difficult for readers to trace their way from start to finish. Mazes start unicursally, at least in their conception.

Nigel Pennick's seminal book on mazes and labyrinths provides a detailed compendium of major examples throughout history and their attendant derivations, narratives, myths, and theories.[19] According to Pennick, some traditions treated the labyrinthine path as a metaphor for life—from perdition to salvation: life's goal seems within your grasp only to be lost and found again along the journey. The unicursal maze's path as traced takes you close to the center, and then further away, only to return eventually to the center. The path also serves as an image of the universe or macrocosm, in particular the movement of the sun as it passes over the four-cornered earth. The sun rises somewhere in the east and sets in the west, but it drifts over the months and seasons in a kind of spiral pattern, particularly pronounced in northern (or southern) latitudes. The maze traces this spiral motion as you follow it, but also as you try to draw the maze.

The unicursal maze provides a simplified diagram of this movement. You start with a square aligned notionally with the points on the compass. You draw arcs connecting points on either side of the square, but asymmetrically. Both the drawing of the maze and its traversal follow this kind of back-and-forth spiral path. In figure 5.2 I show the derivation of the unicursal maze with its origin in the ordinal cross and the four-square grid

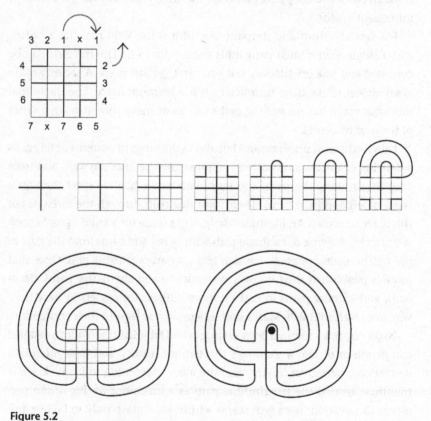

Figure 5.2
Derivation of the unicursal maze with its origin in the ordinal cross and the four-square grid (double tetractys). *Source:* Author.

(double tetractys). Pennick does not emphasize the grid in his derivation diagram,[20] but I think it makes sense to relate the narrative to Vitruvius's account of the layout of the Roman city. Each of the turns that make this a maze occur at the grid.

If you move your body along the path of the maze as a pattern on the ground, either inside a building, outside it, or under it, you are likely to feel dizzy. Hence, the maze is associated with a disordered state—vertigo. So far, I have used the terms *labyrinth* and *maze* interchangeably, and the OED gives them the same meaning. But *maze* is a North Germanic and Scandinavian variant of *mase*, which captures this idea of disorientation: "exhausting labour, nagging, . . . whim, fancy, idle chatter, . . . trouble, bother, . . . to bustle, fuss about, strive, slave away, reiterate, pester, beg, . . . to wear oneself out, . . . to toil, to idle, dawdle, . . . to bask, sun oneself."

My university installed a unicursal labyrinth of paving and gravel in a public park in the middle of the university grounds. It is more complicated than the maze in figure 5.2. During the pandemic lockdown period we were allowed to meet with students outside. I led a column of eighteen students on an impromptu walk tracing the route of the labyrinth. Without theoretical exposition, the most common student banter related to the repetitive and lengthy nature of the journey, and to feeling dizzy. As part of a column of people snaking along the winding path it was apparent that we oscillated into the center and out again until eventually settling on the center. There are no walls to this kind of labyrinth. The character of this space-filling movement had no function other than as a route for exercise and contemplation.

In English, to be amazed is to be bewildered, astonished, perplexed. So, the term *maze* is often associated with a puzzle, involving forking paths, but it also implies exhaustion from the labor and frustration of traversing the length of the path and negotiating all those turns. To reiterate, it is a state that can usefully be described as *vertigo*—a symptom of spinning your body around. In the words of Roger Caillois writing about play and games, such dizziness seeks "to momentarily destroy the stability of perception and inflict a kind of voluptuous panic upon an otherwise lucid mind."[21]

According to Pennick, some traditional dance moves follow the pattern of the unicursal maze, as if to amplify the vertigo effect, and to release the apparent chaos of the heavens. As I've indicated, the traditional unicursal maze structure has the square at its center with tightly packed passageways in concentric rings around that. The concentric rings' maze structure provides a means of relating the fourfold construction of the microcosm, the earth, to the seven planets of the pre-Copernican universe, the seven notes of the octave, and the seven levels of heaven. After all, the method of maze construction delivers seven concentric arcs or circuits. According to another tradition, the center of the labyrinth would be occupied by a young maiden. Any suitor that could navigate the dizzying curves of the labyrinth without stepping off the path would be worthy of the ultimate face-to-face encounter.

To summarize, the maze is a potent metaphor for urban experience. It touches on some of the mythic origins of cities, the significance of the city center, the confusion of winding streets and lanes in old town centers, and the disorientation of encountering a place for the first time. It also contributes to a town's defenses, leading us closer to the purpose and practice of cryptography.

Maze Security

Like the code of a combination lock, the maze also served as a means of confounding access, as a rudimentary security system. The famous maze of Knossos was reputedly such a maze, impeding access to the mythical Minotaur and keeping the beast locked in. In any case, the winding paths would confuse entry and exit. For Pennick, according to this theory, the maze "is a means to arrest an intruder by means of confusion, whilst simultaneously it protects the centre from penetration by any intruder."[22] If there is a tower or summit at the center, then invaders are in full view while traversing the walled maze. They are vulnerable to archers or gunfire. The center also provides a rewarding overview outside the confusion and entrapment of the maze's pathways. The center realizes the aha moment of prospect and clarity, the reward for the tortuous journey required to get there. This unicursal maze has many variants, though it still retains a place as the model of all mazes. It turns up in ancient ornaments, coinage, graffiti, tiling, arrangements of stones, and in paving, as well as built structures and hedged gardens, not to mention secret society symbols and rituals as in Freemasonry. As a link back to the opening sentences in this chapter about a dark wilderness, the maze as an adjunct to a traditional formal garden would be called a *wilderness*, referencing the untamed, unknown, and hostile.

Returning to cities and their codes, as I've shown, the maze derives from the grid and is not so far removed from the contemporary city, its structures, twists, and security systems. We can think of the world as a labyrinth, but it's rare for clients to commission architects, landscape architects, engineers, and others to deliberately create spaces as labyrinths. For most purposes, actual labyrinths are monofunctional and highly inefficient when put to practical uses. They are all threshold, all circulation, at their best with nooks, statuary, benches, and landmarks along the way. They do not provide functional articulation, in other words, rooms with particular purposes. The nesting of passageways prevents views out, or daylight coming in except from above. They slow down transition and require you to negotiate space in a convoluted back-and-forth or circular manner. They obscure and confound navigation. But architects do incorporate lessons from the labyrinth, reference them, and respond to them.[23] The labyrinth serves as an extreme case of the city as system, where circulation takes over. It serves as a metaphor for the city, its infrastructures and communication networks, a theme requiring further investigation.

Cryptopolis

Let us return to the theme of the dark web with which I began this chapter. The idea of the labyrinth conjures up an internal and even a biological space. Elements that belong inside but appear on the outside strike us as grotesque, prime examples of what anthropologist Mary Douglas calls "matter out of place" that contribute to the sense of disgust.[24] Any butcher or spectator of a dissevered corpse would recognize the inner organic world of passages and tubes.[25]

The Persian word for labyrinth is *hezar'to* that translates as "a thousand insides."[26] The urban scholar Somaiyeh Falahat explores the implications of hezar'to as a potent metaphor of the city.[27] Reexamining some of the mazes I described, they appear as nested rooms within rooms—seven in the case of the classical unicursal maze. The passages as drawn resemble entrails—internal organs. Think of these insides as also inside the earth—a multiplicity of interiorities, a proposition that further informs cities and their networks.

Labyrinths occur most frequently under cities. I need hardly repeat the connection between cryptography and the underground crypt in a church or castle (a hidden place) that houses the dead. Architects have certainly paid attention to memorials, monuments, tombs, vaults, mausoleums, and necropolises. The latter are simply cemeteries, though the term *necropolis*, or *nekropolis*, has come to mean something close to an abandoned and ruined city. *Nekropolis* is the final stage in the decline of a city, according to Lewis Mumford, where: "War and famine and disease rack both city and countryside. The physical towns become mere shells."[28]

As reported by Negarestani, the contemporary Iranian architect Mehrdad Iravanian extended the nekropolitan theme. He said, "In order to study architecture, one must first investigate necrocracy."[29] *Necropolis* is in the OED; but *necrocracy* is not. It implies a system of governance where people are ruled by someone who is dead, or perhaps indirectly in a city where people venerate the dead. Such ancestor worship is common enough in autocracies where the dead founder is represented in portraiture or statuary or embalmed in a crypt to inspire pilgrims. The book in which Negarestani quotes this is infused with themes of death. I have referred to his book a few times. He titled it *Cyclonopedia: Complicity with Anonymous Materials*, which among its other characteristics forms an unrelenting compendium, or unstructured encyclopedia of cryptic neologisms.

Negarestani's book is a work of fiction[30] and refers explicitly to code systems, ancient and modern, involving numerals, cuneiform and Arabic script, and geometrical diagrams. It also plays on themes of the crypt, with portmanteau terms such as *cryptogenic, crypto-fractal, cryptomilitary, crypting, decrypting, cryptological, crypto-nihilist, cryptospores, crypto-vermiform, crypto-bureaucratic*, and *cryptic outsidedness*. Cryptography stalks the pages of such literary and urban imaginaries.

Surveil, Ground, Return, and Recur

I have suggested that the maze serves as a metaphor for the city. Visitors lose themselves in the city's streets, corridors, and communication systems. Cities give the appearance of regularity, symmetry, and order, at least as viewed on a map. In his description of cities and places, Borges affirmed that a maze is a house "prodigal in symmetries,"[31] which I take to mean a surplus of symmetries. Probe deeper and you find the city underlaid and permeated by networks, circuits, dead-ends, and short circuits. The best mazes appear ordered and regular from the outside, with slight twists and deviations that deliver the maze's contiguous tortuous space-filling pathways.

There are many similarities between cities and mazes. Here are some other threads connecting them that is relevant to a city's defenses. First is surveillance. The unicursal labyrinth has a center. Following the model of the city as a fort on a hill, that center provides a vantage point. The labyrinth presents as an instrument of surveillance and control. Unlike that leitmotif of the surveillance society, the circular panopticon,[32] the labyrinth assumes movement by those under the gaze of the central tower. In a labyrinth, those under surveillance are not passive but approach the tower, the center. The tower assumes a defensive position. Under this spatial power structure, the journey represents an ascent to a position of control and advantage. By emphasizing movement, the maze metaphor varies the idea of the surveillance society and the city as a site of surveillance. We are being watched while on the move. The all-seeing eye tracks our movements.

The labyrinth in some respects makes maximum use of its enclosing spaces. All of the space is occupied by its passageways. The center of the maze is like the top of a hill. That's where you are rewarded with a view. The center provides the aha moment, where confusion gives way to clarity; enclosure gives way to prospect.

Second, the maze accords with strategies to position the city, any city and its elements, in a relationship with both the heavens and the earth. The classical unicursal maze is derived from the basic architectural grid. In so far as the city is a maze, it participates in ancient legacies relating the city to the microcosm and the macrocosm as I've indicated.

Borges conflated the maze and a residence in a fanciful description in his short story "The Aleph": "In the palace that I imperfectly explored, the architecture had no purpose. There were corridors that led nowhere, unreachably high windows, grandly dramatic doors that opened onto monklike cells or empty shafts, incredible upside-down staircases with upside-down treads and balustrades."[33] Borges echoes this paradoxical thesis in relation to his infinite, labyrinthine Library of Babel: "The Library is a sphere whose exact center is any one of its hexagons and whose circumference is inaccessible."[34] And by library he means "the universe."[35]

Third, the maze speaks to the theme of excursion and return, the practice of the pilgrim, traveler and tourist entering into and departing from the city. A unicursal maze extends the journey from the edge to the center

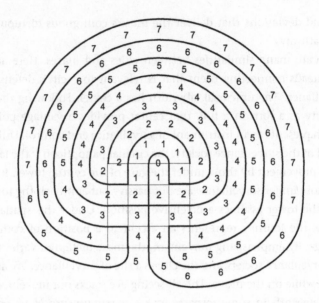

Figure 5.3
Circuits to the center through the maze numbered as steps. *Source:* Author.

in a way that brings you closer at times and then moves you further away. Without that characteristic, the path may as well follow an ascending spiral, which can also be derived from the originary grid—but a spiral is not a maze. The near and far, to-and-fro movement of the classical unicursal maze contributes to the character of the maze journey. For some traditions this movement toward and away from the goal models aspects of life's journey, or any aspirational, hopeful, or dreaded goal-directed experience.[36] Figure 5.3 shows progress around the unicursal maze. The numbers indicate distance in terms of rings from the center. The graph in figure 5.4 shows the journey as an ever-diminishing oscillation toward the center.[37]

Think of the security maze at a crowded airport or the line of students following the university's landscape maze: the encounter and re-encounter with others in line as you loop back and forth toward the goal. Such inadvertent encounters with others are a feature of any maze journey as circulation routes, and people, brush against each other. That reflects the casual and occasional sociability of the city.

Fourth, to elaborate on the idea of a city as a series of nested containers, the cryptographic city is also a *recursive* city. Recursion simply means return. So, a recursive city could be a city that you return to, or that encourages or

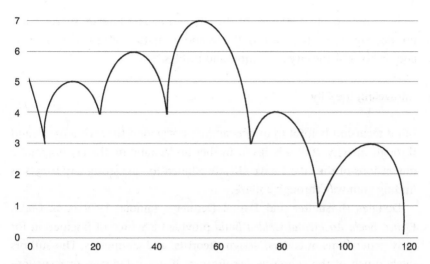

Figure 5.4
Graph of circuit numbers at each step in the progress to the center of the maze showing the to-and-fro movement. The x-axis shows steps on the journey. The y-axis shows the notional distance from the center. *Source:* Author.

requires you to keep coming back—like your hometown or a site of pilgrimage. The metaphor of near and far, of excursion and return applies in many city contexts. Where there is repetition, it's likely there's also recursion—a series of returns. The return may be to a different component that you treat the same as the whole, or the same but at a different scale—for example, a neighborhood grid sits within a district grid that is within a city grid. Recursion applies to geometry, shapes, forms, and arrangements in space, as well as rhythms, sounds, and music. So, the physical aspects of a city fall under the influence not only of repetition, but also of recursion.

In their 2007 book chapter "Imagining the Recursive City" Michael Batty and Andrew Hudson-Smith explain how we can think of the city in terms of recursive structures: "In cities for example, evidence of recursion might be seen in the physical patterns of its development where certain modules repeat themselves in different places, at different times, evidence enough of similar processes at work."[38] They also maintain that recursion shows up in social activity and behavioral patterns: "The way populations in cities behave in space and time, the way individuals, groups, institutions organize their activities, the way economic and social functions determine modes of living and work can all show evidence of such patterning."[39]

In summary, the metaphor of the city as a maze addresses surveillance, positioning within the cosmos, the to-and-fro nature of urban experience, and recursion—the city as a "thousand insides."

Unraveling the City

I find recursion helpful to understanding navigation through a maze, and through the city. It also helps in further understanding the cryptographic dimensions of the city. I have already alluded to navigating city streets as finding your way through a maze.

Douglas Hofstadter and Daniel Dennett's famous 1979 book *Gödel, Escher, Bach: An Eternal Golden Braid* provided a source of fascination for early generations of coders, design theorists, and composers. The authors made much of the cognitive significance, if not paradoxes, of recursion. Think of the M. C. Escher drawing of a hand drawing a hand drawing itself: "The concept is very general. (Stories inside stories, movies inside movies, paintings inside paintings, Russian dolls inside Russian dolls (even parenthetical comments inside parenthetical comments!)–these are just a few of the charms of recursion.)"[40]

Recursion has a precise meaning to mathematicians and computer programmers. There are routines that repeat a procedure over and over, perhaps with variation. That is *iteration*. Hence, a looped routine will draw a series of rectangles to replicate a city grid on a display screen. Recursion is more cunning than iteration. It allows the programmer to define a procedure in terms of the procedure itself. Recursion is particularly useful in algorithms that do navigation and search. Thanks to navigation apps like Google Maps, we take for granted the hidden procedures that find the best route by car from Arbroath to Hampton Court.

Here is a human example of recursion in navigation. Sometimes when someone asks for directions to the railway station, it's helpful to say, "Go to the end of this street, then ask someone there for directions." You are saying: "Go to point A, which is on the network of possible points on the way, and apply the same procedure you did here, which was to ask someone." To get to where you are going, you need to do what you are doing now. More precisely, to go from A to B, select from a set of nodes adjacent to A in the network and go from there. A computer program doesn't usually need to ask for directions but will have a coded form of the city network at its disposal and the recursive definition will search all the possible connections to find the most direct route, recursively.

Most computer programming languages of any power support recursion. Some specialized platforms such as Prolog and Lisp are structured around the idea of recursion. They are designed to keep track of nodes and pathways and are able to backtrack on dead-ends or abortive loops. Hofstadter and Dennett philosophize around such processes in a section of their book they call "Pushing, Popping, and Stacks."[41]

With deeply nested recursion it's hard to get back on track. You can get lost in the city, but also lost while telling a story or writing a book. People tell stories and construct arguments with subplots and digressions. It is fine for stories to wander. But for coherence we expect the storyteller to return to the main point, to rewind the string they just unraveled back into a neat ball, which is the challenge I'm presented with in this chapter, to bring the narrative back to cryptography.

As a digression within a nest of digressions it's interesting to recall that the anonymous Tor Browser as mentioned previously stands for "The onion router" as its relay nodes serve "to encrypt and privatize your data, layer by layer—like an onion"[42] as a series of nested parentheses.

Hofstadter and Dennett write about stories as nested digressions.[43] In a friendly conversation, consultation, or interview the interlocutors may help one another get back on topic by attempting to close off each other's various trailing brackets, or of the progress of the entire conversation. It gets complicated. If the nested parentheticals go too deep then it's hard for both the speaker and the listener to recover, to return to the top or outer-level bracket: nested stories bracketed within stories—as "a thousand insides." Hofstadter and Dennett understand such stories as stacks: "the terms 'push', 'pop', and 'stack' all come from the visual image of cafeteria trays in a stack. There is usually some sort of spring underneath which tends to keep the topmost tray at a constant height, more or less. So when you push a tray onto the stack, it sinks a little-and when you remove a tray from the stack, the stack pops up a little."[44] The customer can only take the top tray.

Stacks are not unique to cafeterias. My browser software maintains a stack of interlinked websites I've visited. Hitting the return arrow button takes me back down the stack. The undo function on most text- and image-processing apps does something similar. A stack is a way for the software to keep track of a series of nested (recursive) computer events.

As I will examine in chapter 7, the stack provides a way of explaining how data gets chained together in the cryptographic "blockchain" as a process in securely transacting digital money. The stack is also a useful metaphor for understanding the spatial organization of cities. Builders stack elements on top of one another, like bricks stored on a construction site. You can't get to the bricks at the bottom of the stack until the bricks above are removed. Though less mobile, urban layers are stacked and accreted on top of one another over time, even down to archaeological strata. Such strata are not as neat as cafeteria trays. It is easy to undercut the order of a stack of bricks, or tunnel through the layers of a city—as it is to circumvent the bracketed elements of a narrative and leave them dangling (to mix metaphors). Precise, algorithmic stacks are the way that computer programs manage navigation through recursive algorithms. A stack is a data table that keeps track of the stages in a recursive process.

Many programming platforms use stack-like data structures to accommodate recursion, and hence provide the means of searching a problem space, such as the best route through the city, that is, exploring a tree of decision points, a branching labyrinth. In chapter 7 I will elaborate on the *blockchain* as a way of securing data, which is a table of data accreted in a stack-like way but is designed so that is impossible to "unstack."

In so far as cities exhibit recursion, they also display characteristics of the stack.[45] Stories are an essential part of our experience of the city. Each nested plot event provides a context for the one nested within it, which in turn informs the events outside the bracketed subnarrative. That accords with the leitmotif of excursion and return I've already mentioned that is popular in theories about play, tourism, and interpretation according to philosopher Joel Weinsheimer: "Essential to play is the freedom of movement to and fro, back and forth, up and down the field—the repeated circular movement of excursion and return that is under control of neither the individual players nor referees but belongs to the playing of the game."[46]

The cryptographic city is recursive, meaning it is a place that invites return—through travel, narration, recollection, and imagination. But the return is never a simple algorithmic return, as in a recursive algorithm. The traveler is transformed in the process, and the city is transformed, like the meanings accrued by nested parentheticals in an engaging story or coherent speech. To simplify, if a city is a labyrinth in its patterns, stories, and recursive processes, then it is also in part cryptographic.

6 The Dissimulated City

Navigating the city presents as a puzzle for many, at least for the first-time visitor. In this respect the cryptographic city is a game space. Many board-games and video games present cities by way of a simulation. Monopoly offers a minimalist rectilinear abstraction of city streets. The early versions of the video game SimCity showed a city laid out on an isometric grid. A simulation of a building is a model of a building as a physical or digital representation created with a computer modeling program or a scan of an actual building (figure 6.1).

Consider the related word *dissimulation*. A dissimulated city is a city that is not as it appears, and in fact conceals what it is. It is in disguise. It offers a fake or feigned appearance. Cities are multilayered, as if wearing a multitude of disguises. Cities typically support a range of communities, each with its own identity. Any community may allege that others are covering over their particular understanding of the city. For the urban critic there is no end to the process of unmasking layers of disguise. In his influential book *Simulacra and Simulation*, the critic Jean Baudrillard pronounced that so many commercial images "dissimulate the fact that there is nothing behind them,"[1] that there is nothing behind the disguise.

Cities, politics, and systems of justice can dissimulate, disguise, and conceal. As well as political critique, the idea of dissimulation informs the functioning of games. Roger Caillois's seminal book *Man, Play and Games* highlights simulation as a key aspect of play, where a person "forgets, disguises, or temporarily sheds his personality in order to feign another,"[2] and enters a "dissimulation of reality and the substitution of a second reality."[3] Dissimulation is a significant aspect of entertainment, games, puzzles, and the cryptographic city.

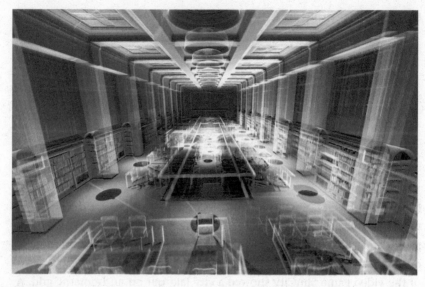

Figure 6.1
An example of 3D dissimulation. LiDAR scan of the National Library of Scotland
Reading Room. *Source:* Asad Khan, www.theentropyproject.com.

Dissimulation is a key aspect of encryption. An encryption is a disguise.
Consider a simple word puzzle: "I am an urban anarxhist." What am I?
The solution is to recognise the errant "x" and replace it with the letter "c"
to disclose "I am an urban anarchist." The trivial challenge in this case is
to detect that something has been substituted, concealed, or *dissimulated*,
disguising an original message. The challenge of decryption is to uncover
that original plain text message. Cryptograms are therefore a species of dissimulation: in other words, disguised messages.

The most cunning methods of disguise are those that conceal that a message is in disguise. Ingenious cryptograms conceal to those outside of the
communication network that there is even a message in play. I will return
to that form of dissimulation in chapter 9.

I want to start in this chapter with the other extreme, where the game
elements of the cryptographic city are made conspicuous and are even
amplified. Urban *gamification* demonstrates the convergence between digital gaming and the city. This discussion also lays some groundwork for an
examination of puzzle solving in cryptography.

Urban Gamification

Gamification involves a range of tactics by which designers encourage reluctant users to adopt novel, unfamiliar, and challenging digital systems and devices. Designers and developers producing human-oriented systems emphasize user experience (UX) design.[4] Citywide UX design comes to the fore with energy-use metering in homes, curbside bike and car rental schemes, accommodation booking, and the panoply of consumer-oriented e-commerce. Contact-tracing apps on smartphones for monitoring and mitigating the spread of infections provide similar UX challenges. Apps for a broad market need to accommodate diverse categories of users, overcome user resistance, and ensure that personal information is secure. There are public and commercial incentives for designers to develop, test, and adapt effective UX design and deploy gamification tactics.

Many of the incentives to lure users into online products and services occur in a competitive commercial environment. Gamification comes across as a commercially motivated tactic for engaging users. It attempts to turn human-computer interaction into a compelling and even addictive experience.[5] By a critical reading, gamification dissimulates the commercial nature of an application, providing a further step toward the commodification of everyday life, reducing urban experience into monetary value or a score, paying little regard to context or diversity of values, and promoting questionable notions of "social credit" ratings.[6] The *Black Mirror* TV program episode titled "Nose Dive" (dir. Joe Wright, 2016) presents a dystopian scenario in which people award one another credit points depending on how they interact with each other, a potent illustration of the dark side of widespread gamification.

Play and Oppression

Life is a game, according to the designation "man the player," *homo ludens*, a term elucidated by the historian Johan Huizinga.[7] The concept of gamification refers to only a part of what being *homo ludens* entails. Even through the limiting lens of instrumentalized, commercialized, and manipulative gamification, the elements of competition, rewards, progression markers, leaderboards, and scores pervade human social organization in urban and work contexts. Formal education relies on gamification as it grades student work and awards degrees. Management structures award pay grades, job titles, and preferred roles on the basis of performance and loyalty. In democracies, people vote for candidates, and are polled, to produce popularity leaderboards. The motivational aspects of scores, rewards, and leaderboards are palpable among spectators and players at sports events, quiz nights, and management retreats.

The movie *The Circle* (dir. James Ponsoldt, 2015) parodies high-tech companies that provide a putatively pleasant, home-like, and leisurely work environment. Such highly successful companies attempt to develop an *esprit de corps* and encourage internal innovation. The movie includes a sequence where the new desk worker handling customer inquiries is introduced to a method for improving customer satisfaction scores. The play context of the work environment is vital for success in the firm. Workers are also required to participate in weekend social events, and to do so voluntarily. Everything the worker does is logged via social media and rated, so managers and teammates know what you are doing, even after work hours.[8] The drama of the movie revolves around the idea that this is all a dissimulation. The jobs are very insecure. The environment is highly competitive and intrusive and will not tolerate criticism from within. Scores, rewards, leaderboards, and teamwork turn out to be a means of exploiting employees and the audience for the company's social media products.

We don't need to concur that ludic, competitive game elements provide the primary motivation for people to do what they do. An article by game expert Scott Nicholson notes that "players engage with games for an exploration of narrative, to make interesting decisions, and to play with other people."[9] He advocates for "meaningful gamification"[10] that encourages players to explore and develop their own individualized goals and motivations.

We could make similar points about education as a gamified experience. Students don't just seek higher grades in their education, but the challenge of the learning experience, the sociability of university life, or the opportunities afforded after graduation. People in work don't just seek higher pay grades, but fresh challenges, and more interesting interactions. Voters in a political race are not motivated to only see their candidate score higher but expect them to bring about some kind of beneficial change when elected.

Combination as Play

In chapter 5 I explored urban combinatorial complexity. To what extent is gamification an exercise in combinations? The answer to that question will advance my case for the cryptographic city. Combinatorial complexity implicates two of Caillois's other characteristics of the game, as described in *Man, Play and Games*. These are the elements of *contest* and *chance*.

Riddles are combinatorial games. I described some riddles in chapter 4. A riddle is a species of game. It is also an exercise in combining elements: a question often delivered in terms of elements permuted in different ways: "What is greater than God, more evil than the devil, the poor have it, the rich need it, and if you eat it you will die?"[11] The unknown entity x is permuted through a selection of phrases "greater than" and "less than" as well as needs and consequences. The answer is "nothing," though the recipient of the riddle and the answer may have expected a more obvious combinatorial variant: "What is less than God, better than the devil, the poor don't have it, the rich don't need it, and to eat it is to live?" That's part of the challenge of the original riddle. The combination doesn't seem quite right.

A multiple-choice quiz is also an exercise in combinatorial complexity. As known to anyone who has tried to score well in such a quiz, you are competing against the power of combinatorial complexity. With ten questions, each offering four alternative answers (a, b, c, or d), there are over one million combinations of responses. There's about a 1 in 500,000 chance of getting half the questions right by answering randomly, in other words, of passing the test if you are only guessing at the answers. A leaderboard is also an exercise in combinations, displaying participants in different orders. Few of us are able to grasp instinctively the scale of combinatorial tasks. But players want to see their team at the top of the list, and they want to know the ordering; in other words, which of the many orderings is the outcome. The very large number of possible orderings makes the game result unpredictable and helps contestants engage in the process. Team configurations interest players as well. There are over ten million ways of assigning fifty players into five groups of ten. So, choosing which group to join similarly engages you in a contest of combinations.

These play elements pervade the cryptographic city, especially if we think of the increasingly gamified elements of urban amenities and services. They also facilitate the operations of cryptography. These are key elements in the arrangement of cities, life in the city, and the city's security systems. The discussion so far prepares us to consider the gaming element of cryptography that involves combinations and puzzles.

Keys and Puzzles

I explained encryption keys in chapter 3. Someone trying to open your four-digit mechanical combination lock would have to try on average 5,000 combinations (10,400/2). That's about two to five hours of work. Computers perform millions of mathematical operations per second. But some sequences of operations still take a long time, measured in minutes, hours, days, or even years. A computer iterating through a potential eight-character password, sequence of passwords, or encryption key by brute force could take several hours. Security software could detect that someone is repeatedly trying different combinations and would shut them out.

One of the ways to foil hackers is to ensure that the effort in working through combinations to break into a system is not worth it. Deploying a central processing unit (CPU) to search through combinations costs time and resources. You need the hardware, network connectivity, power, and time to do the hack before detection, and hackers may need to work on many accounts or passwords at once. A digital database of financial transactions is a common hacking target. In an unprotected financial database hackers could install false transactions that funnel money to their own accounts. Even a legal account holder might try to adjust transactions in their own account. Banks and other custodians of such financial ledgers deploy increasingly sophisticated cryptographic systems to prevent this doctoring of transactions.

As I will examine in chapter 7, in the case of cryptocurrencies (e.g., Bitcoin) there is no centralized data management. The entire ledger of transactions is farmed out to a widely distributed set of computers in the network, with names and other details encrypted so that you can only read details of your own transactions. The security challenge with such a shared ledger is that anyone (a hacker, attacker, or even a regular customer) could plant false transactions. The method for avoiding this is to make the ledger hard to hack, meaning that the codebreaker would have to expend so much CPU resource that it is either impossible or just not worth the effort.

One of the means of making it difficult to effect changes to a distributed database is to set up an arbitrary computational puzzle. A computer involved in verifying legitimate transactions on the ledger would have to solve a combinatorial puzzle that requires an exorbitant amount of processing power. Finding the solution to the computational puzzle draws on CPU time and energy before the solution to the puzzle is stored with the data you are protecting. As transactions are added to the ledger, they get more and more difficult to unpick. Older transactions become increasingly "immutable," practically impossible to change. A hacker would have to expend much more CPU effort than legitimate users to make alterations. The security method is known as "proof of work" to be explained further in the next chapter.

In this chapter I have emphasized the ubiquity of play, cryptography as play and puzzle solving, and hinted at the central role of the puzzle in securing digital transactions. Newer, more powerful computational processors will allow even faster processing speeds. The benefits this will bring include faster iterations through combinations and permutations. That also implies quicker codebreaking, infiltration of security systems, and hacking. In chapter 14 I'll consider the field of quantum computing, which addresses such challenges, with the perennial demand that security systems are designed to be "future proof."[12]

Cryptographic Commons

I have referred already to encryption as a puzzle or game, evident in the history of encryption and encryption manuals published as recreational reading.[13] Encryption and cryptanalysis are hobbies and pastimes for some, evidenced by the high number of explanatory blogs and YouTube clips on the subject. The arts of cryptography appear in stories and films. Even if we lack the patience to solve such puzzles, many audiences enjoy watching other people wrestle with the challenges of encrypting and codebreaking.

Play is sociable. Players compete with other players. They also collaborate with other players in teams, and there are audiences and spectators to cheer on the protagonists. Even solitary play invites spectators, comparisons with other players, commentary, and sociable discussion. By some lights play serves as a means of defusing more perilous antagonism and social disorder. Teams also function to develop and share skills in the game, engender trust, and preserve secret tactics, best moves, and vulnerabilities in the opposition.

As I outlined in the introduction to this book, digital cryptography often attracts the charge that it is an impersonal, instrumental, and rigid process devoid of sentiment or human values. To admit digital cryptography into the realms of human sensibility, some researchers develop cryptographic scenarios that claim to amplify more human-centered and social aspects of cryptography and of keeping secrets. I draw the following example from dyne.org that presents itself as a "non-profit free software foundry."

Consider a fictional cryptographic scenario. A spy (intelligence agent) tells a confidant: "If something bad happens to me then contact Colonel Rodgers at the embassy. He'll know what to do." If the confidant is trustworthy then he or she is equipped to carry out the action as needed. But if some misfortune befalls the confidant then the message is lost. An enemy may coerce or torture the confidant to reveal the secret message. One solution is for the spy to tell several people the secret message. But that increases the risk of betrayal or careless disclosure of the secret message. Another solution is to deliver different parts of the message to a number of people, none of whom has the complete message. On the spy's demise the confidants come together and resolve the puzzle of the message. Had the spy given each member of her trusted circle of friends a fragment of a torn-up letter, or a map, then that would have achieved a similar end as the group came together.[14] The full message is revealed when the pieces come together. That's a familiar mystery story trope: the fragments of a talisman are assembled eventually to become a key to open a secret door to a treasure, or different people bring together fragments of a treasure map. In his novel *The Lost Symbol*, Dan Brown helpfully explains, "Long before talisman had magical connotation, it had another meaning 'completion.' From the Greek *telesma*, meaning 'complete,' a talisman was any object or idea that completed another and made it whole."[15]

Consider the map as talisman scenario but with software—the scenario I described in which people bring together the components of a message in case of a crisis. An application demo at secrets.dyne.org/share provides an interesting model for shared complicity in delivering secret messages. The site contains instructions and a single-entry field. You type a few sentences into the field. For example: "If something bad happens to me then contact Colonel Rodgers at the embassy. He'll know what to do." The program then returns five strings of characters (secrets). These are in code and look like random sequences of characters. The idea is that you email or text-message these five secrets to five of your trusted friends. As the secret is opaque to those friends or anyone else it is not a high-security item. The secret doesn't need to be hidden or invisible. On some signal or other these friends come together. The five secrets serve as a key to unlock your original message. In fact, the entire message is in the secrets. There's nothing stored in a database. The encoding is such that only three of the secrets are required to unlock the message.

The method works even if one or two of the spy's friends are not available to reconstruct the message, or may have lost their secret, or missed the call to action. There's a web page at dyne.org to "combine secrets." Three of your friends paste their string of code in the fields and the message appears intact and in plain text. Only three out of the five friends have to show up to restore the original plain text message, but if even one alphanumeric character out of the three secrets is misplaced, then the message will not be reconstructed. The dyne.org website explains the application of the method: "Secrets can be used to split a secret text into shares to be distributed to friends. When all friends agree, the shares can be combined to retrieve the original secret text, for instance to give consensual access to a lost pin, a password, a list of passwords, a private document or a key to an encrypted volume."[16]

As explained in a 1979 article on this encryption-sharing method,[17] it could be used for authorizing a group decision where the majority is to prevail; for example, more than half the members on a board have to agree before attaching a digital signature to a check. Imagine such a process used in a judicial system, as when an urban planning board has to adjudicate on whether a development is to go ahead, or a planning subcommittee is to release funds for setting up an inquiry. Such applications would fit within overall e-governance strategies for the city.[18]

Though such innovations are delivered through digital technology they present arguably shared, human-centered, sociable approaches to trust. That cryptography can take advantage of trust among groups of agents, human and digitally mediated, resonates with the cooperative aspects of city life and governance. It also suggests applications of digital cryptography that distribute responsibilities for data security.

This chapter began with the idea of simulation, a game function that puts the focus on dissimulation or disguise. I then considered the pros and cons of gamification, a series of tactics for adopting elements of play in delivering practical, functional computer programs and systems. Contest and chance are major elements in play that relate directly to cryptography. They establish extremely difficult puzzles as a way of securing data and confounding would-be codebreakers. Play also has a sociable aspect, an element that some cryptography experts have sought to reintroduce as a means of securing data flows within a community context. Such procedures stand as an aspiration and a motif of the potential for shared participation in digital security. In the next chapter, I deal with the technologies behind cryptocurrencies, which puts the spotlight on the so-called sharing economy, a further aspiration to support sociability in the cryptographic city.

III Crypto-Technics

7 Writing the Block

The city as financial hub entitles urbanists and architects to claim a professional stake in the world of money and finance. In Britain, the *city* is that part of London occupied by the Royal Exchange, the Bank of England, the Stock Exchange, and the headquarters of major companies making up London's financial center, which is still one of the world's major financial hubs. The *city* is therefore synonymous with major financial infrastructure and financial technologies, so-called *fintech*.

Money is the city's lifeblood, vice, and symptom. The cryptographic city is a site of economic interactions. I will demonstrate in this chapter the importance of cryptography in securing monetary transactions as a major element of fintech. That introduces the concept of the *blockchain*, a disruptive technology that claims much promise in the smart city. An explanation of blockchain methods helps illustrate a range of city concepts about encryption, validation, secret communications, and transactions. Whatever the extent of its adoption, the blockchain tests ideas about distribution, sharing, and democratization in the city. The blockchain and the emerging field of *tokenomics* challenge how people think about cities. They also provide an entry point for consolidating the wider role of cryptography across many aspects of city evolution, design, and governance.

Street Life

Lewis Mumford argued that cities have their origins in family and community relationships that are nonmonetary,[1] but the contemporary city depends on financial transactions. Citizens choose among numerous methods for transferring money to pay for goods and services, or to gift money to a charity, friend, or family member. Debit, credit, and cash cards deliver the necessities, habits, and addictions of the modern consumer, aided by cardless surrogates such as mobile payment and digital wallet services (e.g., Apple Pay, Google Pay, M-Pesa). These payment systems rely in turn on multiple layers of encryption to hide transaction information as it courses through networks, thus evading snoops, hackers, leakage, and accidental disclosure. Encryption is almost as vital as money in the digital age. As we have seen, most personal and business communications are encrypted, as are hard drives and files stored in the cloud. Digital encryption hides the content of a message or file by converting it to something that's unreadable by anyone except the designated receiver equipped with translation software and a private decryption key. Encryption and decryption are ubiquitous in digital communications and information processing.

As long as we are digitally connected, secure digital transactions are much more convenient than storing and passing around cash, as in handling coins and notes. For digital transactions we do, however, have to rely on centralized record systems for securely logging the credits and debits of every one of our transactions. With cash you know what you've spent as you see the cache of notes diminish in your wallet. In the case of digital transactions, the records of each transaction are kept by your bank, or a digital wallet service. You also rely on their encryption processes. Such transaction service providers are of course regulated and answerable to governments. So, most of us for much of the time have little trouble trusting our bank. But any taxpayer who has been audited by their government tax department knows that our transaction histories are visible to officials and their systems. You might also be concerned that companies that handle your money can analyze your transactions to profile your spending habits. That data about your habits may also be sold to third parties.[2]

It is only recently that citizens in some countries have become used to walking into shops with no cash in their purses and pockets. But buyers and sellers used to rely on cash transactions, formalized via various methods of recording and receipting. Cash has the advantage that no one but the buyer and seller need know what you are doing with it during the transaction. Cash transactions are person-to-person, sometimes referred to as *peer-to-peer* (P2P) transactions, as discussed in the context of the Tor browser, a computer term implying transactions among social equals, or at least individuals in proximity to one another. In a cash society, you can ask your employer, customer, or bank to pay you in banknotes and coins. Then you exchange the cash for goods from someone else, or indeed give some of your cash away. Needless to say, human societies have been served well by cash transactions. Cash-only transactions hark back to less formal social transactions and are closer to familial social exchanges, bartering, and societies that foreground the exchange of gifts.[3]

Contrary to this story about the benefits of cash, many people now regard cash transactions with some suspicion. Cash-only societies provide opportunities for unregulated and therefore unreliable commerce. The aptly named *Journal of Money Laundering Control* contains numerous articles outlining the pros and cons of various means of exchange, including the prospects, benefits, and risks of cash and cashless societies.[4] Unreceipted cash transactions provide a way of avoiding record keeping, avoiding tax, paying for stolen goods, and paying bribes. That is the dark side of the cash economy. It will become apparent as we progress that digital transactions offer variations on these challenges.

For the time being I'll pursue the line of argument that identifies cash as providing positive benefit to individuals and economies. Urban scholars point to the important relationships between informal (cash-driven) and formal (regulated) aspects of urban living, particularly in developing mega-cities populated by entrepreneurial street sellers, snack vendors, pedicab drivers, and domestic services unable or unwilling to enter into arrangements with banks and the state.

In her influential book *The Death and Life of Great American Cities* in which she extolled the virtues of variety in the way cities form and are planned, geographer Jane Jacobs wrote, "Deliberate street arrangements for vendors can be full of life, attraction and interest, and because of bargains are excellent stimulators of cross-use. Moreover, they can be delightful-looking."[5] The image of the city as vibrant and dynamic invokes the romance of these kinds of person-to-person exchanges, including—I would add—the physicality of cash practices, the immediacy of unmediated exchange of money for goods, and the accompanying habits of browsing, persuading, bargaining, carrying and displaying your purchases, and being seen as active participants in the transactional life of the city. Though many of us can now engage in these routines without handling notes and coins, cash is emblematic of vigorous city living.

Cash Benefits

Cryptocurrencies are an attempt to reestablish the autonomy, flexibility, and confidential nature of person-to-person monetary transactions. As I have shown, encryption is ubiquitous in online activity anyway, but cryptocurrencies involve a special class of online transaction mechanism. Others have followed, but the first major, successfully engineered cryptocurrency was Bitcoin that emerged in 2009 around the time of the banking crisis. At that time trust in banks was at a low ebb. According to the seminal paper that launched Bitcoin, a cryptocurrency is "an electronic payment system based on cryptographic proof instead of trust, allowing any two willing parties to transact directly with each other without the need for a trusted third party."[6] Normally you would trust your bank or other transaction service provider to keep accurate and secure records of your digital transactions. In the case of cryptocurrencies, the trust is relegated to a series of smart algorithms distributed across a digital network.

A cryptocurrency such as Bitcoin is digital money that purportedly carries some of the benefits of cash. You can buy things with it, give it away, invest it, and stash it without involving a bank. But unlike cash, there is no physical paper or coinage. Students in one of our classes design banknotes as a graphic exercise, using graphic software. But they can't use those notes as currency. Were we to design our own banknote or make a PDF of a scanned real banknote and email it to a friend, we would soon find that it carried no value; in fact, you would still have your own copy of the file, which means it could be reproduced and sent to someone else. Like real paper money, digital money has to be scarce. When you give some of it away, then you should no longer have what you had. As a further illustration of the need for scarcity, my local fitness center used to send me a PDF of a free visitor pass when I answered an online customer satisfaction survey. That free visitor pass served as currency of a kind. I would approach the reception counter with a friend and a printed copy of the coupon, and they would let the friend enter for free. They stopped sending out free passes when it became obvious that customers could print any number of copies of their free pass and use it more than once. It wasn't scarce, so it quickly lost its value as a means of rewarding and regulating customer activity.

Simulated, digital versions of paper money are redundant anyway. Our money with the bank or digital transaction platform is just a series of texts and numbers in a database. As you receive or spend money the balance goes up or down, and we trust the bank to keep an accurate and secure record. As I have suggested, digital money is for people who don't want banks or transaction services to manage, monitor, and profit unduly from their transactions. People want the benefits of cash but without the physical paper or coinage. It ought to be possible to process digital money as records of transactions on a digital database.

A Shared Ledger

When desktop computers were still a novelty, I supplemented my PhD study by designing and maintaining an accounting system for the residential college where I lived. It was a Victor 9000/Sirius desktop computer running dBASE II. With no real instinct for accounting, the first conceptual hurdle to overcome was figuring out how to create a database for each resident's transactions, that is, the expenses they incurred and their payments. That wasn't necessary of course. A system to record transactions with the residents could be stored as one big database rather than as a separate database for each resident. Transactions would include amounts billed for accommodation, extra meals, sports fees, payments, and refunds. As long as each transaction record clearly identified a unique ID for each resident, the date, what the transaction was for, and the amount of money, then the computer could generate monthly reports that only showed the transactions for that individual. That way, the residents could be sent specific billing information each month. As they paid their bill then the accountant would enter that payment as yet another transaction, a credit, in the ledger. Each time we printed an account statement for a resident a balance would appear at the bottom of the page showing what was owed, or in credit. It was simple enough to devise an algorithm to quickly total all the transactions for any resident to deliver their current balance.

This is basic accounting, an update on double-entry bookkeeping, one of the progenitors of Enlightenment rationality as I outlined in chapter 2. I mentioned accounting systems here to help explain what is meant by a ledger, and how it can accommodate an arbitrary number of users and their transactions. The database of all transactions was the ledger, which mirrored the ruled pages in the college accountant's paper-based bookkeeping system, which my system would eventually replace. There need only be one single ledger, even though it records the transactions of hundreds or thousands of customers, clients, or residents.

My system was stored on one computer in one place, managed by the accountant and me. The accountant needed access to every transaction. In fact, she or her assistant, entered the transactions herself as the invoices and cheques came in. Between us we played the role of a bank as record keeper. The challenge for a cryptocurrency is to maintain such a ledger. But any customer needs access to their own transaction records at their own computer or smartphone. That personal, user-level access is common enough now with digital money flows, online banking, cash cards, and so on. The next challenge for a cryptocurrency is to keep people from tampering with the ledger, eliminating their own debts or entering false payments. Banks of course are specialists in such security. The college accountant and I kept our little system and its database securely locked in an office, and we were the only ones with the room key and computer password to edit the transaction records.

The idea of cryptocurrencies is to avoid such a centralized storage facility, and the main challenge of cryptocurrencies is to obviate the risk of tampering. In the case of decentralized cryptocurrencies, the solution to the security problem is paradoxical. To keep the ledger secure, a cryptocurrency system creates and maintains multiple copies of the ledger and distributes the copies to a large number of computers in a network. That's the ledger of all transactions, not just ones belonging to any particular customer. To continue the analogy with our in-house accounting system, that would be like distributing electronic copies of the ledger file to a trusted collection of college residents and requiring them or their computers to update the ledger each time a new transaction comes into the main office. They would have to keep comparing their versions of the entire ledger to see that nothing has been altered among the previous transactions. That is an implausible and time-consuming challenge for human beings but can be easily automated now with networked computers.

In the case of cryptocurrencies, software compares distributed copies of the ledger to one another in case someone tries to double-spend an amount of money or modify the record. In fact, nothing gets recorded in the ledger unless computers on the network agree it should be there, and the distributed versions of the ledger are compared to each other to make sure nothing has been changed. This is a task for dedicated computer processors on the distributed cryptocurrency network whose job it is to validate transactions. In the case of Bitcoin there can be around ten thousand or more active nodes on the network, labeled as "fully-validating nodes."[7] The state of the distributed system fluctuates as computers equipped to undertake the validating procedure come online or go offline.

Once the cryptocurrency software is distributed for this process, then it should take care of the ledger. The software that runs the cryptocurrency ledger is maintained collaboratively as an open-source project. Regular users wanting to make transactions most likely encounter the currency through third-party platforms that provide digital wallet services such as Luno, Bitpay, BRD. These platforms operate as intermediaries providing access to the cryptocurrency ledger, but none of them monitor or maintain the ledger or monitor what the transactions are for. In fact, the encryption protocols are designed so that neither a human being nor a computer algorithm could read or deduce who made the transaction, who was the recipient, or what it was for. Users of digital wallets gain access to their own account of transactions via a unique private encryption key (a kind of passcode). If they lose the key, then they lose access to their transaction records. No one can redeem that information for them. The value of the account is lost.

In my simple in-house college example, I think there were three reasons the accountant was happy for me to develop a bespoke accounting system rather than buy one from a commercial supplier. First, such commercial systems existed, but the printouts looked like impersonal computer printouts as opposed to the personalized letters the residents were used to. Second, my system was designed to mirror as closely as possible the college accountant's paper-based bookkeeping system. Third, I also designed my system to mirror her ability to correct the ledger if she or her assistant made a mistake. I thought that if the accountant were to over- or under-charge someone, or mistype a payment, then she would enter another transaction (e.g., a refund) that corrects the mistake. The new transaction is labeled as an adjustment. But that would appear on the account sent to the resident as an adjustment arguably creating an unprofessional impression. So, I did what any sensible accounting system would *not* allow—I provided a user-friendly interface that let the accountant enter the record system and adjust

the ledger to correct it. In the wrong hands, there was little to stop someone
with access to hack the ledger or insert false payments. Everyone involved
in this case, however, happened to be extremely trustworthy: nothing went
awry, and in any case external auditors would be performing annual checks.

But a distributed ledger cannot allow alterations to transactions whether
to correct a mispayment or to extort. How does a cryptocurrency system dis-
allow alterations to the ledger, especially as it is so widely distributed? What
happens if the different versions of the ledger don't match up? Here, analogies with conventional systems for keeping accounts fail. The best I can
think of is the banal practice of writing transactions into a paper ledger in
ink rather than pencil. Then they can't be erased or altered, or if they were
then the alteration would be obvious to an auditor looking at the ledger.

Securing Transactions

For a cryptocurrency, the way to secure each transaction in the ledger is to
provide some method of locking it down so that it cannot be accessed or
changed. Transaction records are secured by a form of cryptographic coding
known as *hashing* introduced in chapter 4 as the translation of data into a
short piece of random-looking code. The way I developed an appreciation
of the method was to spend time on a website that has an empty field into
which you type or paste an item of text. The website then returns the corresponding hash string. The site run by xorbin.com implements the SHA256
function, the name given to one of several standard secure hash functions.[8]

A hash string is a kind of signature of the transaction, a 64-character-long
string of arbitrary-looking alphanumeric characters. That hash string can
be created for any body of text, of whatever length. For example, you
could run a standard hash algorithm to create a hash of the text of one of
Shakespeare's sonnets or the entire set of ordinances for a city. That sixty-
four-character hash string will look nothing like the text from which it is
derived, and it is impossible to reconstruct the sonnet or the ordinance
from the hash string. But if even one character in the original document is
changed, or a comma is out of place, then the hash string of that version
will look different, and any pattern-matching software could easily detect
that the file has been altered. If you email the text of the document and its
hash string to someone, they should then be able to reproduce the same
hash string from the body of text you sent. If the hash string is different,

that means the document has been altered. Comparing hash strings is a way of checking if a document has been tampered with or corrupted in transit. The process is automated of course. The algorithm that does the comparison can't detect what has been changed in the file, but it could certainly signal that the file needs to be rejected. The same technique is applied to transactions in a ledger.

At this stage it is worth noting that the algorithms in the cryptocurrency system check and secure transactions in real time. For efficiency the transactions are processed as groups of transactions. The Bitcoin digital money system processes and secures around five hundred transactions at a time as they come into the system and irrespective of who made those transactions or where they were made. This collection of validated transactions is called a "block," from which the terminology *blockchain* derives.[9] The shared and ordered ledger and its verification apparatus is called the *blockchain*. The agreement to accept a block of transactions is reached by algorithms running at the node computers in charge of validating blocks of transactions.

The method of agreement is another conceptual sticking point to most people new to the working methods of cryptocurrencies. I alluded to the method in chapter 6. I will elaborate here pending further explanation in chapter 11.

The algorithms at the node computers, meaning the fully validating nodes, try to outcompete one another in solving an arbitrary numerical puzzle that requires brute-force computation to solve. The node that finds the solution to the puzzle first receives some digital money as a reward, and the answer to the puzzle is planted into the block of transactions as a record that the most recent transactions are verified. Individuals and companies that dedicate processing time to securing the blocks are called "validators." For most cryptocurrency systems they are also referred to as "miners" as they automatically generate digital money as a reward for their role in keeping the currency secure. They effectively generate money for themselves, which in turn increases the supply of Bitcoins. Their digital reward is simply a transaction recorded on the ledger credited to the miner. The miners keep what they generate, though they incur the considerable cost of the hardware and electricity to run their processors solving the numerical puzzle.

Anyone so motivated can become a Bitcoin miner. You don't need approval or a license. Sites such as www.bitcoinmining.com provide instructions on how to become a miner.[10] In the early years of Bitcoin mining individuals with powerful desktop computers could download the software and configure their home computer to contribute automatically to the contest to validate blocks of transactions. As the competition between miners has increased, so has the demand for more efficient processing. Now, there are specialized microprocessors that you plug into a home computer arrangement. For individual miners, the usual practice is to connect to a mining pool. That is a consortium of computer owners who share processing power and distribute any profits from Bitcoin mining among themselves. By most reckonings, home-made Bitcoin operations will now barely make enough to cover the cost of the electricity consumed by always-on Bitcoin mining hardware.[11] Stories about crypto-mining circulate amongst enterprising students. Ethereum (ETH) is a cryptocurrency related to Bitcoin. One of our student informants reported about a friend: "A software engineer working in London at the time, was mining Ethereum (ETH) back in 2017, making enough money to offset his increased electricity bills and save the excess ETH." The mining challenge has since migrated to companies with the resources to run large-scale specialized processing farms in places where electricity is cheapest. This constraint has lured cryptocurrency mining to countries such as China. As of 2021, China ran 65 percent of all Bitcoin mining operations, though its activity in the area has varied with the rise and fall in energy costs and responses to government policies.[12]

I'll put aside details of the validation process for the time being. Most of the transaction information in the Bitcoin ledger is in plain sight and can be read on a computer screen on web pages, for example, at blockchain.com/explorer. You can see the amounts of money being transacted in each block on the ledger, but information about who is spending or receiving the money and what it is for are encrypted as a series of arbitrary-looking hash strings. The visibility of this ledger is of value to researchers assessing the performance of the Bitcoin ledger's operations. Though it is in plain sight, the confidentiality of individual transactions is maintained by the cryptographic protocols. No humans are involved in the process of verifying transactions. That is done by algorithms. I'll explain more about the arcane procedures of the blockchain as I progress, but for the time being it's worth drawing attention to the implications of cryptocurrencies for the cryptographic city. That is the subject of chapter 8.

8 Bitcoin Cities

If you are in any doubt about the relevance of cryptocurrency to the city, then consider Bitcoin City. In 2021, the president of El Salvador declared Bitcoin legal tender. So businesses had to accept the digital currency as well as U.S. dollars. The country had already abandoned its own currency, the colón, in 2001. Even more interesting from an urban perspective is the president's adoption of a new idea for a city, Bitcoin City.[1] The electricity grid of the new city is to be powered by geothermal energy from the Conchagua stratovolcano. The city plan is designed by Mexican architect Fernando Romero. The city will sit beside the volcano and follows a common utopian circular plan form. It will be ecologically sustainable, and the Bitcoin "mining" will also be powered by electricity generated from geothermal energy.

The International Monetary Fund (IMF) has urged El Salvador to abandon the use of Bitcoin as national currency, stressing the "large risks associated with the use of Bitcoin on financial stability, financial integrity, and consumer protection."[2] Vendors are reluctant to exchange Bitcoin for goods and services, not least due to the volatility of the currency and the time it takes for the Bitcoin blockchain to validate a transaction. Currencies rise and fall relative to one another, encouraging speculation as investors trade optimally one currency for another. For all the benefits of peer-to-peer, private, cash-like digital transactions, the implementation of the blockchain so far has encouraged widespread speculation, with subsequent impact on urban life.

Money Markets

Cryptocurrencies are ripe for speculation. People buy Bitcoin online with money from their own bank accounts, or money transferred from some other source. They can then spend their Bitcoin with vendors who accept the currency. You can buy Bitcoin by registering with a website. That then interfaces with a Bitcoin wallet on your laptop or smartphone. As a risk-averse investor my own forays into cryptocurrency were modest. In July 2017 I bought £100 worth of Bitcoin. One Bitcoin was then worth about $6,000.[3] That gave me about 0.02 Bitcoins (to six decimal places). My digital wallet resides on my smartphone, which I took with me on a holiday in Ukraine as a backup in case I had trouble with my debit card. I was mainly interested in testing how easily I could transact in Bitcoin. As it happened, no vendor there was able or willing to accept payment in Bitcoin. So, it was not much use to me for buying goods. After I returned home, I encouraged a colleague to open a Bitcoin account and transferred £20 worth of Bitcoin to his wallet. That was simple enough and we organized the transfer directly via our smartphone wallets. Around the same time, I kept my eye on the value of my crypto assets on currency markets. It was rising rapidly. My £100 less £20 rose in value over the space of five months to £510. As a cautious trader I decided to sell and transferred £480 of the currency to my regular bank account, leaving £50 behind to reap the rewards of a further rise. Soon after that the value of Bitcoin plummeted. So, I got out in time, and I was ahead on the deal. Had I borrowed £100,000 from a bank and spent it on Bitcoin during that period of reckless speculation then I would have accrued £480,000, enough to buy a London bedsit, and the government would have done well from the tax on my capital gain. Other punters have been more successful. One student informant told me: "Back then I invested mainly in ETH and Litecoin, using Kraken and Coinbase, and I also had to get a bit creative to cash out part of the tokens. The purchase and sell prices were around ten times up. It was enough to help me buy an ice cream shop on a Greek island with a friend, but in hindsight it would have been considerably more if I had waited until 2021 to sell."

Needless to say, the value of cryptocurrencies relative to real (fiat) currency is unstable. Rather than spend their cryptocurrency on goods, people tend to buy and sell it in the expectation that it might rise in value. Since its release, the value of Bitcoin rose rapidly. It created Bitcoin millionaires, and many people of course lost money when its value dropped.

In the current state of the technology in 2022, very few merchants accept Bitcoin. They do not have the means to process it, they can't be sure of its value relative to fiat currency, and although transactions are processed in real time, they are not instantaneous. It can take from ten minutes to a day for the consensus among miners to release Bitcoin to the vendor. That may suit online purchases but is unfeasible for buying goods in a shop, when the vendor needs to know the transaction is secured before you walk out of the door. At the time of writing, cryptocurrencies don't yet outcompete the convenience of chip and PIN debit and cash cards, and mobile phone transfer accounts such as Kenya's M-PESA.[4]

There are costs to maintaining fiat currency: printing notes, stamping out coins, banking infrastructures, the materials and energy that have gone into office buildings, main street banks, and iron safes. But cryptocurrencies also entail costs. I've already alluded to the cost of the electricity to run a Bitcoin mining operation. The costs of running a blockchain are not trivial, and they are additional to the costs of banking infrastructures that already exist. Verification of cryptocurrency transactions is costly to the environment.[5] Competition among miners to verify blocks of transactions and thereby earn Bitcoin has escalated the computer processing needed to compete. The process requires hardware and power and generates heat as an unsustainable resource burden on the environment. The idea of "sustainable cryptocurrency mining"[6] that uses energy from sustainable and renewable energy sources, as claimed for Bitcoin City, remains a topic of research. The mining effort also concentrates the computational effort to corporations and countries that can afford to support these processes. After all, the verification process by the validators (miners) only delivers value in so far as it sustains the blockchain, and the cost of that escalates due to competition. This concentration of economic power has political implications and arguably reconcentrates power among the few that can afford to run a mining operation, rather than the many. That is antithetical to the original cryptocurrency ethos, at least as promoted.

Digital currencies, as part of the sharing economy, ostensibly enable peer-to-peer exchange independent of centralized control hierarchies. They carry the benefits and vices of cash economies. As with black and grey economies, you can exchange digital money for goods and services without being traced or having to declare income to the tax department.[7] But digital money is also corporatized. My Bitcoin wallet is connected to a network node or hub in Luxembourg. As indicated in chapter 7, Bitcoin "mining" is dominated by big companies with CPU farms in China, and large "pools" of profit-sharing miners that work against the idea of a decentralized system. Where large profits are involved, it is common for successful grassroots enterprises and initiatives such as cryptocurrencies to succumb eventually to the pressures of scale. Either they grow to become big corporations or consortia, or big firms take them over.

Alternative blockchain models are emerging. The incentives are high for researchers and developers to devise methods that obviate the exorbitant energy, CPU burden, and electricity costs of validating blocks of transactions, especially in light of the climate crisis and attempts to meet carbon reduction targets. A substantial part of the cost resides with the *proof of work* (PoW) process by which validating nodes on a blockchain compete to solve a numerical puzzle, the solution to which is then circulated to all other verifying nodes and embedded into the latest block of transactions. The winning miner in this contest is rewarded with some cryptocurrency. It is a competitive process with potentially high financial rewards, producing an incentive structure that according to a review article by Fahad Saleh has "triggered a computational arms race among PoW validators. That arms race manifests in PoW blockchains expending an exorbitant level of energy."[8]

One solution that's offered to the arms race is to develop methods that assign the validation procedure randomly to computers on the blockchain network. The randomness is weighted toward those that are already most heavily invested in the cryptocurrency of that blockchain. The validation process is known as "proof of stake" (PoS). On the face of it, a randomly selected validating computer has no incentive to reject illegal blocks that come its way. The method also implies that control of the blockchain will reside with the richest participants, those with the highest stake. Saleh's article shows how neither tendency diminishes the functioning of the blockchain, nor does it cause a "rich-get-richer" effect.[9] The developers of the Ethereum (ETH) blockchain that supports Ether cryptocurrency claim to have switched Ethereum to a PoS validation method in September 2022, incurring the risks of adjusting a system that already listed transactions worth millions of dollars.[10]

Smart City Contracts

Blockchain technology intervenes in the narrative ecology of the city aided further by its support for smart contracts.[11] As with a line in a bank statement, blockchain-enabled cryptocurrencies such as Bitcoin store lines in a shared ledger indicating payer, payee, date, amount, and the goods or services to be exchanged. Money transactions provide the prime application and the motivation for developing blockchain technology. But it has other uses. Instead of a text line indicating the product being exchanged, that line could include a piece of computer code that implements some consequential action in response to a trigger event. The Ethereum blockchain platform (ethereum.org) supports the idea of transactions as code. An Ethereum ledger line can contain a "smart contract" as an active piece of computer code that carries out some actions as part of a blockchain transaction.[12] The signatories to the contract are anonymous, but the code, the contract, is visible to anyone on the public ledger. Distributed knowledge that parties have entered into an agreement helps keep them accountable.

Writing about the role of blockchains in a sharing economy, Arun Sundararajan explains the smart contract method: "The smart contract protocol can specify, as computer code, terms under which certain obligations are fulfilled, and can execute actions like sending a payment or deactivating a file once there is evidence of the contract's terms being fulfilled."[13] He lists some of the benefits, or at least changes in business practices, that may follow the adoption of smart contracts: "Smart contracts are autonomous if after they are finalized, the initiating agents theoretically never need to have contact again. Smart contracts are also self-sufficient to the extent that they are able to marshal their own resources. Finally, smart contracts are decentralized; they are distributed across network nodes rather than residing in a centralized location, and are self-executing."[14] Smart contracts need involve no one but the parties who have signed up to them. The record of their existence resides in the Ethereum blockchain.

Smart contracts require a means of interfacing with the world in which they are to operate.[15] An "oracle" is an item of hardware or software that channels information to the smart contract on the blockchain: that the goods have been received, the warranty has been invoked, the goods have been sold on to another buyer, or that other conditions of the contract have been met.

The idea of the smart city aligns well with the concept of smart contracts. The professional work practices of city governors, architects, and others are populated with formal contracts. A 2020 study showed that video games, currency exchange, and gambling exhibit greatest smart contract activity,

but researchers and blockchain specialists have identified property contracts as a major area of application.[16] Chainlinklabs.com asserts the relevance of smart contracts in real estate, with examples such as defining
conditions under which property ownership can be transferred, securing
collateral for loans, triggering rental payments, and transferring real estate
ownership based on predefined conditions.[17]

Researchers have examined the application of smart contracts to building
information modeling (BIM). BIM is a widely accepted platform for designing, documenting, and managing buildings and construction projects
involving integration across all specialties, and is supported by established
computer-aided design suppliers such as Autodesk and Bentley Systems. A
smart contract item could be used to describe and visualize a parameterized
building component (e.g., door, wall, staircase, street furniture), with attendant actions, constraints, and rules for use. The code in the contract could
also indicate how agents can deploy, exchange, copy, reproduce, or dispose
of such digital assets. BIM researchers Shojaei Alireza and colleagues assert,
"Building Information Modeling (BIM) due to its data-intensive nature and
the level of details presented in an appropriate model is an excellent way
to tie different sections of the work to a smart contract."[18] At the time of
writing, these applications hold promise but are as yet untested in the BIM
marketplace.[19]

Smart contracts bring into relief challenges presented by codified and
automated contracts. Try as we might to be exact, contracts are ambiguous,
contingent instruments that fit particular situations. The automation of a
complex contract implies that the contract code accounts for every even-
tuality in the domain of application. Furthermore, peer-to-peer contracts
will have to draw on expertise, or at least contract templates put together
by knowledgeable experts. But secure contracts that automatically facilitate
the actions they prescribe is an alluring prospect for the smart city.

Blockchain Urban Metaphors

I began chapter 7 with cryptocurrencies and drifted to the underlying technology of the blockchain, a powerful tool for securing information flows via
encryption methods. How does this narrative relate to the city? The digital
world has long brought metaphors to bear on how we think of cities—as

flows of data, networks, circuits, grids, and an "Internet of Things," as if cities are made of bits, memory (RAM), sensors, actuators, and with communication systems, inputs, outputs, and operating systems. The metaphor of the blockchain is potent. It provides analogues with city living, not least as we think of the data-intensive smart city, the overlay of integrated and responsive digital infrastructures that draw on big data streams from mobile apps, sensor networks, social media feeds, and transport information, to make buildings and transport systems more responsive to changing conditions.[20] In so far as we credit these expectations, we might assume that such infrastructures will operate under centralized control. Blockchain technology claims the potential for an alternative, localized, grassroots, and democratized dimension to the smart city.

Continuing with the metaphor, as in the case of El Salvador's Bitcoin City, "mining" is suggestive of volcanoes and geothermal energy. Blockchain processes also demonstrate the geological metaphor of stratification. In a blockchain, data is layered in a time-ordered sequence as a kind of stack. The oldest is the deepest, with layers of data cemented by computationally byzantine verification procedures. Cities are like that in some respects. As well as physical stratification, people talk about cities as layers and accretions of memories, some of which are inscribed in the fabric of a place. We want to peel back the layers and watch translucent layers interact as they get scoured and replaced. But at the same time, some like to think of a city's memory strata as immutable. Try as they might, those who would like to hack the past find resistance from the accretion of embedded layers.

The blockchain also draws the urban scholar's attention to the challenge of validation within authority structures. In a blockchain, the validity of one item of data depends on the validity of another. Some hierarchies are like that, as are the structures of city governance. But so are informal relationships in communities. The good pupil inherits the respect accorded to her teacher, who in turn is deemed a good citizen by local shopkeepers, who are in turn validated by the respect they gain from their customer base. This is a conservative model. In architecture as in city governance, we build on the credibility of the achievements of others.

I mentioned the process of mining to secure the blockchain. A hacker would need to expend substantially more energy than the miners to access and change the blockchain, decoding and peeling her way through the layers of the blockchain. As outlined already, securing the blockchain is called "proof of work." This is the process by which nodes in the blockchain network contribute CPU time and effort to solve the arbitrary but extremely difficult cryptographic puzzle, the solution to which gets printed into the blockchain to confirm the validity of a block of transactions. This is not entirely alien to social functions in the city. From a semiotic perspective, to expend effort is to indicate a commitment and to validate your intentions. Think of the circumstance where political campaigners go from door to door to persuade would-be voters. It is not always the reasoning that persuades people, but the fact that someone braved the weather and spent the fuel and body energy in an effort to come and talk to the potential voter at home. Would-be persuaders are even more persuasive when they invest effort in something—preferably related to their cause.

To expend effort is to prove that something is of value. Putting in the effort shows the strength of your conviction. There is an argument here justifying otherwise unprofitable civic projects: follies, memorial statues, pyramids, and public art. That someone cared enough to spend and risk valuable resources, money, design effort, good will, and reputation on a building, artwork, ornamentation, or infrastructure project strikes any city visitor as a statement that the city has values. There is care there.

Making something visible, or at least accessible in a public way contributes to trust. That is one of the attributes of the distributed ledger idea in blockchain technology. The structure and its content are visible to anyone who wants to inspect them. Transparency is a watchword of good governance. It is a way of keeping people honest.[21] As with digital surveillance, the blockchain idea amplifies such vehicles of trust. But blockchain transactions are purposefully peer-to-peer, with the most private parts of the transaction encrypted, as long as you don't lose the key.

Having read up on El Salvador's Bitcoin City proposal, I see that cryptocurrency is there in the symbols, narrative, and form of the city diagram, though as yet I don't detect anything about land use, exchange, or commerce that draws on ideas of unregulated peer-to-peer exchange. The sharp boundary to the city's circular geometry also works against the symbolism of sharing and openness. Nor is there a display of pipes, reservoirs, turbines, and steam valves that speak to sustainability and celebrate the thermal energy that is to power Bitcoin City.

Many contradictions come to light as we probe the city via the lens of the blockchain. A city as distributed ledger would lay everything out to be viewed, used, modified, and accessed. But, like my Bitcoin wallet, access is only granted to those with the private decryption key. There are many parallels here with the distribution of software and other online assets. You need the key to unlock the features you have paid for. As any architect knows who has had to draw up a key schedule, a building is a system of locks and keys. So is a city—a matrix of locks, keys, vaults, hidden spaces, security doors, cameras, contactless sensors, keypads, and passcodes—fixed and mobile. Under the blockchain metaphor, cities reveal themselves as hyper-encrypted, and the security of the smart city depends on that.

The Politics of Cryptocurrency

I hope it is obvious by now that cryptocurrencies entail an ideology of decentralization that is attractive to many. The concept of decentralized commerce sounds suitably progressive and democratic. The ideology appeals to extremes on both the right and left of politics. For those on the right, cryptocurrencies support the idea of unregulated trade, individualism, small government, and self-sufficiency. The ideology harbors an undercurrent of right-wing, anti-establishment views as well as self-reliance and mistrust of the state. For researcher David Golumbia, digital currencies "emerge from the profoundly ideological and overtly conspiratorial anti-Central Bank rhetoric propagated by the extremist right in the U.S."[22] By this reading, cryptocurrencies arguably are money for anti-establishment "preppers" suspicious of the "deep state" and other targets of alt-right opprobrium.

On the left, cryptocurrencies suggest a means to greater democracy, people power, and a break from large corporations and financial institutions (i.e., a break from capitalism). The banking crisis of 2008 eroded the trust of many in banks and financial institutions. That cryptocurrencies are "decentralized" suggests the democratization of money and empowering people over institutions. Like informal cash exchange, cryptocurrencies may encourage certain kinds of grassroots development. Many people in developing countries are mistrustful of banks, or don't have the resources to borrow or benefit from what banking offers. Cryptocurrencies promise a means of empowering the "unbanked."[23]

On the negative side, unscrupulous operators also see such communities as fair game for scams. The ubiquitous reach of the Internet delivers, promotes, and amplifies the operations of financial products and provides a vehicle for influence campaigns that exaggerate claims about cryptocurrencies. Cryptocurrency is mysterious to the average consumer, but tantalizing. There is plenty of scope for companies to deliver explanations that both clarify and confuse, deliberately or unintentionally. It is also easy to insert false claims into a narrative dressed up as a cryptocurrency project.

Bitcoin was the first successful cryptocurrency and now there are others, also known as "altcoin,"[24] including Ether, XRP, Litecoin, Zcash, Monero, and Dogecoin. Apart from their apparent utility, several factors amplify enthusiasm for new cryptocurrencies. The story of Bitcoin lingers as evidence that cryptocurrencies provide an opportunity to acquire wealth quickly. Some startups in the blockchain world recruit investors by selling "initial coin offers" (ICOs) as a way to raise startup funds. Investors buy the cryptocurrency at a favorable rate in the hope that their holdings will eventually increase in value.[25]

As well as enabling legitimate cryptocurrency schemes, these factors create a climate rich with opportunities for scams, get-rich-quick schemes, fraud, and fake systems that exploit the appeal of cryptocurrency narratives concerning, for example, wealth, decentralization, self-sufficiency, community, opportunity, education.

Crypto Scams

I'll conclude this chapter with a flamboyant illustration of a cyber scam. Cybercrimes extend to scams and faux cryptocurrency schemes. The "onecoin" scandal was such an illicit operation.

I first heard about onecoin through a BBC Podcast called *The Missing Cryptoqueen* by journalist Jamie Bartlett and producer Georgia Catt who investigated the scheme and the damage it has wrought on individual lives. As I listened to the first episode of the podcast, I thought I was hearing a mockumentary, or a mystery story in the form of a documentary, and perhaps an elaborate allegory of political scam culture about how easily people get duped into cults and adopt reckless chants at rallies. But onecoin is real—or at least it is a real fake, a dissimulation that exists. Onecoin (and its company OneLife Network Ltd) effectively runs a pyramid selling scheme—a Ponzi scheme. You pay to join the network and get some payback as you recruit others to join. Those new network members in turn get a cut as they recruit more people. The more people you recruit and the more they manage to recruit others, the more money returns to you. The scheme preys on myths of multiplication. You multiply your financial investment in the scheme by recruiting others who in turn multiply returns to you, exponentially. Money percolates upward through the pyramid. The organizers and early members inevitably get most of the money. Of course, the market is finite, and later adopters run out of people they can coerce into buying into the scheme.

I recall my pre-digital school days when such pyramid schemes circulated by word of mouth. They seemed to involve posting money in an envelope back to someone who recruited you to join the scheme. You then had to recruit others who would send you money. Parents and teachers of course warned against participating. So, I never fully understood what it entailed, or got rich. More charitably, this approach is sometimes termed *network marketing*, relying on a dedicated group of individuals and their social contacts. In the case of onecoin you pay to join in fiat currency (i.e., £ or $), and your income from the scheme accrues in onecoin, a supposed cryptocurrency. Members can see the value of onecoin rise relative to fiat currency on the onecoin website—or so it seems.

As it happens, there's nowhere to exchange onecoin for fiat money. You cannot deposit or exchange it at a bank. Merchants in on the scam only sell products if you pay in a combination of onecoin and fiat currency. Most important, the supposed value of the currency is only what the organizers say it is worth, and they keep inflating it to make members feel as though they are getting richer—and keep hoping that one day their virtual riches will convert to actual money. The tragedy for many people who buy into the scheme is that once they realize this won't happen, the cost, anxiety, blame, and guilt filters through their social network, straining relationships among friends and loved ones recruited to the scheme.

A telling video available on YouTube shows the onecoin founder, Ruja Ignatova, inspiring fans by demonstrating how much better onecoin is than Bitcoin, the promise it holds of future riches for its members, the strength of the community of members to determine how the currency operates, the speed and convenience of transactions (unlike Bitcoin), the benefits of its delivery of learning materials, and so on. As a demonstration of how robust and community-centered the organization is, Ignatova explains how OneLife Network has just doubled the value of each person's onecoin account on the ledger. The crowd cheers.

The OneLife Network also sells education packages, for real money, that teach you about cryptocurrencies. It is a learning community after all. Here is the punchline: the BBC says that the founder made more than $4 billion for herself. She has not been seen since 2017.

In spite of this disappearance and exposure in the media, as of 2021 the scam continued to recruit. There are other scams based on cryptocurrencies. There's dagcoin; I also discovered trumpcoin online.[26] I won't link to their websites lest that add to their connectedness and their claims to legitimacy. Investigators say there is in fact no blockchain to support or verify transactions in onecoin. If onecoin is anything at all useful it is merely a centralized database of transactions. Such Ponzi schemes trade in the claimed benefits of multiplication: riches that accrue from your own participation in a hierarchy of mass-produced wealth.

There are other scams that implicate cryptocurrencies. The circulation of false narratives in social media about adversaries is a deception that hurts the target, but garners support from the believers and hence funds via merchandising and donations. The "Luther Blissett Project" was an invention of an anonymous activist Italian art collective in the 1990s. Art commentator Eddy Frankel cites this collective as one of the inspirations for the QAnon conspiracy movement—social conspiracy co-creative performance as art project.[27] Other commentators, such as Izabella Kaminska, have linked the methods of QAnon to cryptocurrency.[28] The invocation of the letter "Q" and its made-up stories are supposedly sourced in the U.S. Intelligence Service. An interesting article in the *Financial Times* blog pages in 2020 argued, "Cryptic messaging, puzzle-solving and anonymity/pseudonymity feature prominently in both systems."[29] By "both systems" the author controversially binds QAnon conspiracy theories to Bitcoin. Art, conspiracy theories, performance art, and cryptography feed off inventive combinations, however edifying or perverse—a crypto-combinatorics as it were.

At the start of this chapter, I took El Salvador's initiative to create a Bitcoin City as an opportunity to discuss speculation in money markets, the risks and uncertainties of cryptocurrencies. I reviewed the idea

of blockchain-based smart contracts as instruments for automated transactions in the city. The blockchain idea serves as a way of framing our
understanding of the city, suggesting various metaphors of city structure
and living. We looked at the political affordances of cryptocurrencies, and
finally the susceptibility of the city to crypto scams. I will return to some of
the technical aspects of the blockchain in subsequent chapters. My discussion so far has been based largely on cryptography as a means of managing
texts and numbers. In the next chapter I turn to the impact of cryptography
on the city as a visual arena.

9   Images in the City

Pictures are vital elements of the smart city. In their book *The New Urban Aesthetic*, Degen and Rose note the role of computer-generated images (CGIs) in manufacturing and promoting the impression of a place: "CGIs are now a pervasive part of the marketing of new urban developments, large and small, in very many cities across the world."[1] Cities compete against one another with images of spectacular buildings and development proposals. Pictorial representations of urban settings are prominent in promotional material, photo-sharing platforms, personal collections, display screens on mobile devices, and mounted on walls and building facades. Pictures conceal as well as reveal, and are important instruments in the functioning of the cryptographic city.

Hidden Pictures

Children's picture books encourage readers to identify elements and clues concealed within images.[2] John Soane's house conceals pictures behind hinged wall panels (figure 1.2). Artists can treat surfaces so as to reveal hidden images when viewed from a particular position and angle (figure 9.1). I recall the craze in the 1990s for *Magic Eye* posters and books. People would gaze and squint at seemingly random multicolored patterns to discern hidden 3D images of dolphins, elephants, temples, and spaceships. Such *autostereogrammetry* was a variation on what I experienced as a child, as I gazed at my bedroom wallpaper that at times seemed to be on a different plane to the walls. Our chain-link garden fence presented a similar disruption to depth perception.

The presentation of full 3D imagery from just a single picture requires computer manipulation of patterns of pixels that repeat across the plane of

Figure 9.1
Concentric circles installed as thin painted aluminium sheets on the walls and towers of Carcassonne. Designed by Felice Varini (2018). Seen from other viewing positions the installation appears as swaths of arbitrary yellow patches across walls and roofs. *Source:* Rayints via Shutterstock.

the image. The process is created from a depth map, which is a grayscale pixel image of a 3D object, such as a temple. An algorithm maps this data onto a repeating pattern. It adjusts pixels on adjacent repeats in the pattern according to the information on the depth map. The variable convergence effect provides the depth cues, and we see the 3D shape. At the time of their popularity these images reminded us of the remarkable capacity of the human visual system to take in details (pixels) presented to each eye from the same source and to match them so as to signal depth to present a 3D illusion. The techniques assume full visual acuity of course. There is no coloring; the repeated patterns drape across the surface of the 3D model. You may see ghosts and echoes within the imagery, and if you allow your eyes to lock onto certain repeating patterns then the 3D imagery fragments and multiplies. As with head-mounted displays (HMDs) and 3D cinema, it is not yet possible to simulate the way the human eye adjusts focus according to depth cues. There's no parallax either. You can't see around the virtual 3D objects by moving your head.

Those of us involved in computer graphics had trouble identifying roles for autostereogrammetry techniques, especially in architecture. They don't seem to offer glasses-free 3D cinema, though a YouTube video of the Young Rival music group shows the musicians moving in 3d animation through visual white noise.[3] There are many other animated examples of autostereogrammetry.[4]

So-called "cerebral cryptography" adopts these and other techniques to incorporate the capability of the human cognitive system to view secret images and messages unaided.[5] The principles can be demonstrated simply. In the 1990s, text email (without pictures) was the main channel for person-to-person communication. Mimicking early text printer graphics, arrays of evenly spaced characters could serve as simple pixels inside an email. Imaginative emailers followed this method to adorn their messages with emojis and ASCII pictures. Among such text-based imagery you can find *ASCII stereograms*, arrangements of symbols on a page that when viewed in a certain way reveal 3D pictures. A short online article by computer scientist Jonathan Bowen demonstrates the practice.[6] ASCII stereograms offer a low-tech means of hiding messages.

A seminal 1995 article by Moni Naor and Adi Shamir outlined how they could secure printed text, handwritten notes, and pictures such that they "can be decoded directly by the human visual system."[7] The challenge was to transmit short, singular messages. Even though computers might be required to generate the encrypted message (ciphertext), its decryption could be accomplished by purely visual means, such as laying images photocopied onto transparent sheets of acetate: "The basic model consists of a printed page of ciphertext (which can be sent by mail or faxed) and a printed transparency (which serves as a secret key). The original cleartext is revealed by placing the transparency with the key over the page with the ciphertext, even though each one of them is indistinguishable from random noise. . . . Due to its simplicity, the system can be used by anyone without any knowledge of cryptography and without performing any cryptographic computations."[8] Their 1995 article has over three thousand citations, and it was still being referenced in 2022. So the concepts are significant.

As a variation on the method, a 2017 article by Petrauskiene and Saunoriene discusses how secret messages can be hidden and revealed by subjecting images to physical vibrations: "Secret information is encoded into a single stochastic moiré grating, which is fixed onto the surface of the vibrating structure. It is shown that the secret can be visually decoded if the cover image oscillates according to a chaotic law."[9] Secret messages may be hidden within moiré patterns, 3D magic eye, and other innocuous imagery.[10]

The methods of concealment I have described so far in this chapter are satisfyingly physical, a proposition that appeals to anyone interested in the physical built environment. But they are mostly curiosities. In digital communications, encryption that is useful needs to be fast, scalable, and reliable. The practical uses of ASCII stereograms, cerebral cryptography, and vibrating moiré grids are unclear. But to hide secret messages in plain sight brings cryptography into the realm of the human visual system, and hence the senses. It also brings the challenges of cryptography into alignment with optical illusions. Whatever their practicality, these methods are useful in shedding light on questions of perception and reality. Not least, knowledge about how pictures can conceal things challenges trust in the visual sense. Distrust of pictures has a history that impinges on the perception of urban settings.

Iconophobia

In chapter 2 I described the architect Leon Battista Alberti's important development of cryptography. Alberti's affinity with writing was evident in other ways as well. In spite of his obvious engagement in the visual cultures of architecture and painting, Alberti expressed mistrust of pictures, especially in the reproduction of books. By pictures, I mean diagrams, illustrations, drawings, and other visual representations. To produce multiple copies of books, scribes and copyists would repeat sequences of symbols, of letters and numbers. Variations in the form of those letters made little difference to the message. Reproducing and disseminating texts prior to the widespread adoption of the printing press was challenging, but pictures were even more difficult to reproduce. Letters, words, and numbers repeated by many hands over several generations could withstand idiosyncratic variations of hand and quill, but inaccuracies in drawing would compound over repeated reproduction. Where they existed at all, illustrations were frequently removed in the copying process. For example, there are no surviving pictures from Vitruvius's first-century *Ten Books on Architecture*,[11] if in fact he ever produced pictures.

Alberti was aware of the possibilities of the printing press, and that seemed to have influenced his thinking about architecture. But he was also wedded to the constraints of the old methods for reproducing texts. Only after his death was his *The Art of Building*[12] printed on movable-type printing presses.

As a progeny of the pre-press era, Alberti was guarded about the inevitable inaccuracies that would result from reproducing illustrations. So, he avoided illustrations, even though the subject matter of his book *The Art of Building* was well suited to visual illustrations. Subsequent printed editions were populated with diagrams introduced by others. A study by Mario Carpo and Francesco Furlan provides a helpful explanation of Alberti's apparent "iconophobia."[13] Alberti followed the tradition by which authors relied on textual descriptions to ensure the longevity of the (virtual) images they wished to convey, which in turn relied on the imagination of the reader, who would construct mental images from the author's vivid prose—an inevitable aspect of telling and listening to stories. In any case, writing and drawing involve different skill sets, and not all writers can draw well enough.

Not all drawings need to be accurate, except when attempting to convey ratios and proportions, and plans and elevations of buildings and their details, as Alberti wanted to do. Cartography provides similar challenges. As a significant illustration of his rational avoidance of pictures, Alberti produced a map of Rome, its boundaries, river, and landmarks, not as a drawing, but as coordinates. He thought successive copying could reproduce the table of numbers more faithfully than a drawing. The map could be reconstructed accurately from the coordinates by anyone who wished to see it.

In that pre-cartesian era, Alberti adopted a means of defining locations within a city not on a two-dimensional right-angled grid, but following the methods of radial geometry, meaning, locations in space defined by angles and distances from a single origin point in the city, the top of the Capitoline Hill.[14] Alberti provided a detailed description of the method he used to create the map, and the tool needed to reconstruct it. The latter is a circle drawn on a sheet of paper (he called it a "horizon") marked out in forty-eight numbered segments, with four further divisions to each segment and a distance scale (the "spoke") that pivots at its origin from the center of the circle as an arm on a clock face. The spoke was to mark out distances from the origin point.

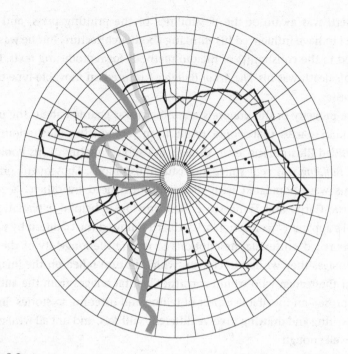

Figure 9.2
Reconstruction of Alberti's map of Rome as it appeared in *Descriptio urbis Romae*, by architect Luigi Vagnetti (1915–1980) in *Lo studio di Roma negli scritti albertiani* (1974). The outline of the modern city is shown in grey. Redrawn by the author from Jessica Maier, *Rome Measured and Imagined: Early Modern Maps of the Eternal City* (Oxford: Oxford University Press, 2015).

I attempted in the previous paragraph to describe without pictures how a particular technical apparatus appears and operates. That is another response to iconophobia. In the current proliferation of images and uncertainties about intellectual property, it is sometimes more convenient to describe in words than incur the costs and risks of copying pictures from other sources, though I have succumbed with my own version of the map so produced in figure 9.2.

Alberti's plan illustrates further his apparent interest in codes. After all, without the benefit of a computer, his tables of numbers did not allow anyone to actually see the city plan without some specialized labor translating those numbers to a drawing. I think his method of illustration fits nicely with his interest in ciphers. It is a vivid illustration of Rome rendered in code as a prototypical "cryptographic city." The tools and the method served to provide pictorial access to an otherwise opaque table of numbers.

City Grids

In contemporary terms, Alberti's method of diagramming the city was a vector representation—a series of points joined by lines. Much two-dimensional and three-dimensional information about buildings, land-scapes, and cities takes the form of point and line information. After all, solid objects have edges bounding surfaces, and the conventions of plans, elevations, and perspectives rely on mathematical projections from this kind of data. Early computerized productions of maps and architectural drawings would control a physical pen moving across a sheet of paper. Screen displays that worked with vectors would aim streams of electrons to the back of a phosphor screen to draw lines visible to the operator as luminous green marks across the screen surface.

Cathode ray tube technology in early computer displays converted lines and surfaces to colored pixels. Pictures produced on the current line of HDR (high dynamic range) displays are often imperceptible as regular arrays of pixels. Thought of as arrangements of grid cells, pixel images have a privileged relationship with the built environment, due not least to the priority of the grid in the way streets, buildings, and open spaces are conceived and laid out.

Much architecture is about the grid, though building grids don't have to be as regularly spaced as pixels on a display screen. Arrays of grid cells, or pixels, provide a useful means of depicting spaces, as plans and elevations of buildings, zones, regions, and city blocks. Floor plans that don't easily match regularly sized grid cells can be stretched, twisted, and simplified to conform. As an example, figure 9.3 shows the plan outline of the Helsinki railway station by Eliel Saarinen adjusted to fit a regular pixel grid. This crude outline can be stored efficiently as two blocks of information: data about the grid line spacings and the pixel values of the grid cells. In this case, "1" means the cell is enclosed by the building, "0" is external to the building.

At much higher resolution, arrays of pixels are the dominant means of depicting space in digital imagery. A cluster of pixels can stand in for a building in plan, a city block, region, a landform, landmark, field, forest, or country. Aerial and satellite photography delivers pixel images at various

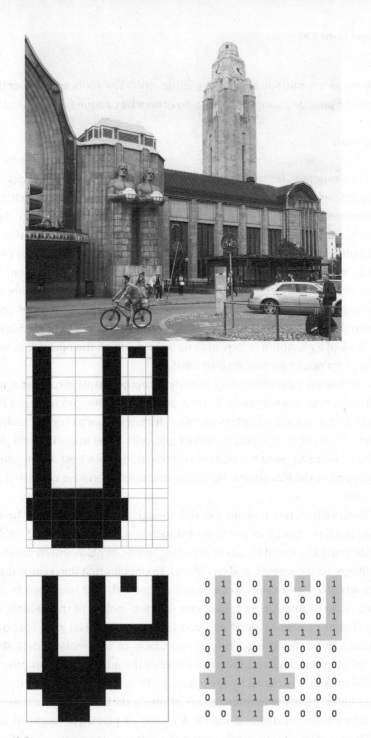

Figure 9.3
Simplified plan of the Helsinki railway station by Eliel Saarinen adjusted to fit a regular pixel grid and turned into binary code. Photograph and drawings by the author.

resolutions, scales, and values within and beyond the visible color spectrum. Pixelated display screens are, in any case, ubiquitous in cities. As well as CGI promotional animations, there are giant billboards, digital watches, and display screens that present arrays of luminous cellular pixel units that are part of the fabric of cities.

Pixel images have the advantage that they lend themselves to rapid calculation. Algorithms can derive characteristics of a space from pixel values. Aerial and satellite remote sensing typically records light spectrum information across regular grids over the landscape. Through stereopsis, successive image frames from aerial cameras reveal elevation data for cityscapes and landforms. Algorithms derive three-dimensional terrain models from elevation data to provide perspective views, or derive information about slope, exposure to sun, view sheds, heat maps, and contours.

Pixel grids at the scale of remote sensing or snapshots taken on a digital camera lend themselves to image manipulation at the scale of the pixel. Averaging color values blurs images. The reverse process sharpens an image. Image editors such as Adobe Photoshop, smartphone photo editors, Instagram, and Snapchat provide graphic filters for stretching, twisting, superimposing, and otherwise manipulating images.[15]

Standard digital image manipulations sharpen and blur images at boundaries. These are processes that occur in urban spaces as well, as the distinction between city and countryside *blurs* at the edges, and as the new freeway serves as a *sharp* barrier between neighborhoods. The way we describe cities draws on metaphors encouraged by drawing, painting, photography, and image manipulation. These descriptions in turn influence urban processes.

There is a long tradition of such exchanges between visual image production and what happens on the ground. The architectural theorists Colin Rowe and Fred Koetter's book *Collage City* comes to mind in emphasizing the confluence between image manipulation and the form and life of cities.[16] The idea of the urban *palimpsest* is also relevant. Parchments used to be reused and written over in a succession of writings and erasures. New writing demanded that the ink would be scraped from obsolete documents and the parchment reused, sometimes leaving a residue of the previous text or imagery behind. As I discussed in relation to blockchain metaphors in chapter 8, cities present such successions of layering as older structures decay or are removed. Any city will contain remnants of what existed before at various levels of exposure or concealment.

I am deliberately conflating the idea of the grid cell as a device of spatial organization at the scale of the city, with that of the pixel on a display screen. That serves to further align the spatial organization of cities with cryptography. As I will show, pixel images are subject to cryptographic processes. As this book is about the ubiquity of cryptography in the city, I'm bound to focus on the pixel image. It is one way of accessing the spatial aspects of encryption. Pixel images can be scaled and compressed. They can also contain secret signs, such as water marks, as well as other pictures and hidden messages.

Hiding Pictures inside Pictures

Designers, digital artists, and cartographers position drawings, pictures, and models over one another and manipulate the transparency and prominence of those layers. The layer in a building plan that shows the position of steel reinforcement in the concrete floor slab may be of interest to a structural engineer, but of less interest to the designer or subcontractor concerned with positioning the office partitions. Digital design platforms will enable the designer to switch off or hide layers depending on the task at hand. Practitioners in urban architecture, engineering, and construction (AEC) exchange digital drawings in which the layers are explicit and follow standards, such as the layers associated with partitions, cabling, pipework, and so on.[17]

Those layers may include both vector and pixel representations. It is the layering of pixel images that interests me here. Cities are much more than their pictorial, pixelized representations, but the fact that pixel images can hide information may reveal something about the way cities are represented and organized. What I describe here is in some respects analogous to an urban palimpsest, though conducted with digital precision.

The practice of hiding images is *steganography*, which entails hiding one picture (a secret or data) inside another (a host or carrier) image. It involves a different set of techniques than drawing overlays that can be switched on or off. With steganography, the hidden image (or layer) is not an isolable part of an image file.

Steganography usually involves hiding a message in an innocuous picture so that no one but the recipient knows it contains a secret message. That is different from passing around encrypted files, which already advertise to receiving software the fact that there is something hidden.

As with cryptography in general, there are several reasons someone might want to hide content via steganography. The hidden image could serve as an invisible digital watermark. If someone copies a picture and distributes it without the owner's permission, then the owner can detect and recover the watermark to prove ownership. Hiding one image inside another also serves as a means of passing secret messages around. The secret picture could be a top-secret map or other confidential data. The carrier image is in full sight. Only someone with the right algorithm can recover the secret data.

Bad actors can use steganography to pass around compromising and illegal images. An extortionist might conceal compromising pictures in public display screens or digital posters. With the right key, these will suddenly reveal their secret content. Thankfully, I've not heard of this happening, but such images could conceivably be printed out as posters, to be revealed by special software on mobile phone cameras. Interest in steganography and how to defeat it are growing.[18] Some of the techniques and principles are of interest as we explore the cryptographic city.

Least Significant Bits

Steganography exploits the redundancy of much image information. It also exploits the limit of the human faculty to detect very small changes in image quality. It is worth probing this method of hiding images for analogues with architectural production.

Computer files are made up of numbers and symbols in binary code as bit strings of 0s and 1s. The usual method of hiding one picture inside another is to exploit the characteristics of binary number representations. Most full-color images have at least 255 shades each of red, green, and blue (RGB). The binary value of 255 is 11111111. It is an 8-bit number: the minimum value of a pixel is 0 and the maximum is $2^8-1=255$. For an 8-bit color picture, every pixel will be assigned a value for RGB in that range. To keep things simple I'll just consider a grayscale picture, meaning, pixel values that are in the range of black to white 0–255. Grayscale images made up of 8-bit color are of good quality. Sixteen-bit grayscale is even better. That is $2^{16}-1=65,535$ shades of gray. Large numbers of gradations allow for subtle color variations. This is particularly important where there are large areas of sky or other subtly graded surfaces. The fewer bits there are, the more likely the pictures are to show banding across large areas that have a subtle grada-tion of color. Four-bit grayscale imagery is still acceptable to most viewers. A 4-bit grayscale picture has a range of values 0 to 1111 in binary. That is 0 to 15 in decimal numbers, or 0 to $2^4-1=15$ shades of gray.

Binary calculations are slow for human beings, but extremely fast and efficient for computers. In fact, to reduce the range of color values of a picture from 8-bit color to 4-bit color it is only necessary to remove the last four bits of each binary pixel value, that is, to cut the bit string in half, and discard the half to the right in the bit string.

An occasional misplaced 1 or 0 in the pixel value of the trailing, right-most bits of an 8-bit string will be scarcely noticed by a human eye. These trailing bits are called the "least significant bits" (LSBs). Pixel images are "noise tolerant," in other words, we humans are so good at identifying the patterns in a picture as a whole that we gloss over minor discrepancies in color values. Steganography exploits the human tolerance for noise in pic-ture data. Importantly, you can also pack extra information into the trailing bits while retaining the integrity of the original picture.

The first 4 bits are the most important for establishing the color value of a pixel. So, the last 4 bits could be used for a second image. Adding extrane-ous 1s and 0s in place of the last 4 bits degrades the host's 0 to 255 range of grays toward 0 to 15 shades of gray and introduces some scarcely visible noise. Importantly, the hidden image would be invisible as it provides only a subtle variation in the color values of the original. The recovered image will be at a lower quality (color range) than in its original state. The first half of the secret bit string gets appended to the first half of the host string. If a pixel in the host image has the value 10101101 and the value of the same pixel in the secret image is 11100111 then the combined pixel value will be 10101110. If that process were reversed, the secret would be visible and contain the host as the hidden image.

It is clear that the "least significant bits," meaning, the bits at the low end of a binary string of the kind used to specify color values for a pixel, could contain not just pictures but also any kind of information, code, instructions, or malware (stegomalware). The idea that you can code hid-den information in the least significant bits of a file of binary numbers, in particular a graphical image file, is fascinating from the point of view of cryptography. Once the last 4 bits of an 8-bit string are surrendered to hidden content then it is apparent that the order of those bits could be

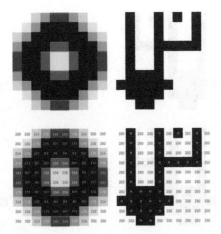

Figure 9.4
A carrier image on the left and the image to be concealed on the right. The bottom row of images includes the grayscale pixel values. *Source:* Author.

changed as part of the encrypted coding, and the ordering of those LSBs could vary across the whole image according to some formula, as long as that formula could be replayed by the algorithm that recovers the hidden image (figures 9.4 and 9.5).

Challenges of Concealment

Steganographic techniques are designed to conceal pictures, to conceal the fact that there is anything concealed, and to evade hacking or modifying the hidden message. Any cryptographic technique faces challenges from a codebreaker. To obviate detection, the size of the file containing the host and its secret image should be the same as the host on its own. So, the secret image should not be hosted in some metadata attached to the image file or extra bits of data. The graphic display program should display both the carrier and its secret content. The secret image should all be in the pixels, as if it is entirely in the surface, in plain view. The secret image should be undetectable by the human eye. The carrier plus its hidden picture should also conceal the fact that there is a secret picture. If no one is to be aware there is something hidden, then the encoding of the hidden picture will have to evade software that detects excessive noise in the image, or that indicates a compromised color range: the file size may indicate 8-bit color,

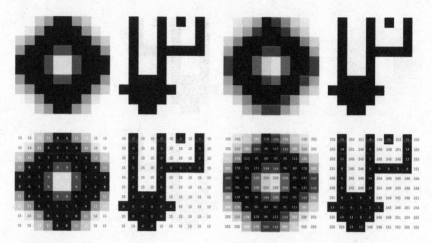

Figure 9.5

Demonstration of two images combined by adjusting the LSB (least significant bit) of the carrier image. The first image in the row is the carrier image at 4-bit gray; the second image is the 4-bit image to be hidden, the third is the two combined in 8-bit gray, the fourth image shows the hidden image recovered, with little disruption to the image quality. The bottom row includes the grayscale pixel values. *Source:* Author.

but it is apparent that what is on display is only at 4 bits. That would indicate something is hidden in the image data. To enhance the hiddenness of the hidden image it could be further encrypted in various ways, requiring a private decryption key to expose the secret content.

Other precautions are needed. Hackers could substitute a different hidden picture or watermark for the one planted by the message sender. If a hacker tries to scrub or distort the original hidden picture then that should be apparent in the hacked image. If the secret image is to serve as a watermark to prove that it is your image then there needs to be a method of ensuring the integrity of that watermark, and that a hacker cannot overwrite it with another watermark. The best steganographic techniques will withstand the application of compression algorithms, scale reduction, and other manipulations that erase or degrade the hidden image.

It is sobering to think that these battles take place over the invisible content of pictures, that the visual field, digital imagery, and screen culture generally could harbor hidden images. Presumably hackers, photo-sharing platforms, or camera software could hijack your own digital photographs and insert hidden images, and without your awareness. If it is possible to hide pictures inside other pictures, then text can also be secreted in a similar manner exploiting properties of bit strings, in other words, the least significant bits, and the noise tolerance of image files.[19]

There is an architectural connection here. In so far as architecture deals in structures, functions, and surfaces, steganography in architecture contributes to the dissimulation of surface. There are also interesting correlates to superfluous and redundant content in urban form. The redundancy of ornament is a common theme in art, design, and architecture. When it comes to structure, form, substance, and function the ornamentation is arguably "the least significant bit." It is the part that can be adjusted or even removed without jeopardizing the integrity of the whole. Yet it is capable of delivering substantial significance on behalf of the structure. I am thinking of traditional stone masonry, where the masons were given rein to shape and adorn minor parts of the building. Brackets, bosses, capstones, and rainwater spouts could be carved as flowers, crests, tools, portraits, animals, and grotesques. We might say these adornments introduced a layer of coded, and even secret, meanings.

To stretch the association further, consider the Austrian architect Adolf Loos's book *Ornament and Crime*. It is a celebrated polemic against the superfluity of ornament in the modern industrial age. Ornament is not innocent: "not only is ornament produced by criminals but also a crime is committed through the fact that ornament inflicts serious injury on people's health, on the national budget and hence on cultural evolution."[20] I don't think he had in mind that rogue architects would use ornamentation to conceal hidden messages, but his book suggests an association between ornament as "least significant" and aberrant insertions into space.

Concepts of urban layering, palimpsest, and collage imply that cities are made of varied intensities of meaning with differing social and physical affordances. Some of these layers are "least significant" according to who is doing the reading. Urban habitats for wildlife, plants, and microorganisms present as interstitial, contiguous, and mappable but "least significant" layers of the city. Further analysis reveals that they are essential contributors to air quality, carbon capture, water management, and biodiversity. These "bits" show up not just in managed parks and gardens but as mosses, lichens, fungi, insect colonies, and microorganisms on roof surfaces, derelict buildings, and underused urban non-places. By this steganographic reading, it is as if urban analysts must strip away from their remote sensing image data the vivid and conspicuous to reveal these subtle strata in an urban biosemiotics.

Returning to pictures, QR codes and barcodes are also graphical images and can be secreted into photographs, printed packaging, advertising graphics and brand logos using steganographic techniques. Texts concealed within pictures could include a password to a bank account, a private decryption key for a file, or a trigger to activate some malware.[21] The hidden message could be any arbitrary string of code, including a hash code. By my reckoning, a raw picture file of 3000×3000 pixels with 0 to 255 shades of RGB could hide a file of over three million characters, meaning, three books. It would be less if you allow for error checking, compression, and further encryption.

The secretive bits don't need to be distributed evenly across the whole picture array. Steganography can work in concert with compression techniques. JPEG compression reduces the color depth in those parts of the image where contrasts between adjacent pixels are high. Some of the redundant color depth could be recruited for hidden message bits. Messages can be hidden at the edges where subtle color variations go unnoticed. I will review some of the implications of image compression in chapter 10.

Crypto Art

It is easy to see this kind of image manipulation as underhanded, motivated by theft and extortion. But artists, designers, and innovators are also interested in the creative potential of media manipulation and in protecting their production. This is a growing field within the purview of digital art, and more specifically crypto art. I will conclude this chapter on digital images with an account of this emerging field.

Authors of books commonly need to incorporate pictures that belong to libraries, museums, archives, or other individuals. I have paid for the right to reproduce certain digital images from www.gettyimages.co.uk and shutterstock.com for book illustrations, but that's not the same as purchasing a work of art. I mentioned watermarking as a way of staking a claim on a drawing, photograph, or other kind of digital image. Techniques based on LSB encoding can provide this kind of security. But asserting, maintaining, and policing ownership is notoriously difficult in the case of digital assets.

Figure 9.6
Trompe l'oeil art mural titled *Mur des Canuts* in Lyon, France, by artists of Lyon (1987). Photograph by Pierre Jean Durieu via Shutterstock.

In the current climate, high-profile digital artworks present as static or moving digital pictures such as animated GIFs, though the category includes any digital asset such as videos, sound files, music scores, computer code, 3D computer models, and mixed-media performances.

Hidden art exists as an art subgenre, often subsumed under the themes of optical illusion and anamorphism. The related subgenres of paradoxical art, faux perspective, and op art (optical art) also exist (figure 9.6). A web search for cryptographic art turns up a substantial subgenre. Steganographic art is an emerging genre, and implies concealing one artwork inside another using encryption. But *crypto art* is a term now reserved for artworks that are bought, sold, and authenticated on a blockchain.[22] At the time of writing, crypto art focuses on mechanisms for securing rights to artworks.

Artists and dealers attach a certificate of authenticity (COA) to a physical artwork to verify their claim that it is an original and not a copy. Anyone can forge a COA, so many creators log the certification in some centralized registry, ledger, or database, as happens in the case of birth certificates, university degree testamurs, and patent certificates. If you don't trust the piece

Figure 9.7
Work by voxel artist Mari K (https://madmaraca.art) and sold as an NFT. The work
is titled "Emiris—The Forgotten City" and references Italo Calvino's *Invisible Cities*.
Printed with permission from the artist.

of paper, you can always look up the certification at the relevant trusted
registry.

Non-fungible tokens (NFTs) are an attempt to replicate this operation on
a blockchain. NFTs are digital equivalents of certificates of authentication.
They are digital COAs attached to an artwork and registered on a blockchain. At the time of writing the most common artworks certified in this
way are digital, screen-based works, though some designers have sold NFTs
for 3D virtual architectural works (figure 9.7).

You can purchase digital artworks on sites such as the superrare.co digital
art trading and auctioning platform. You need money in Ether (ETH), the
Ethereum blockchain's cryptocurrency, to make a purchase. What can a

purchaser do with their crypto art? It is screen-based art, but the purchaser does not have the automatic right to publish it in other media, nor is the artist prevented by contract from continuing to display it online. A helpful blog by Hexeosis examines the various conditions for the exchange of crypto art.[23]

In March 2021, Christie's auction house sold the NFT of a two-dimensional pixel image at online auction for nearly $70 million.[24] The picture is still reproducible, and the creator retains the right to do other things with it, short of selling it again, though the buyer may sell it on in a secondary market. Some crypto art entrepreneurs have reduced their art to an image that appears about as insubstantial in size as an NFT. In their own right, such certification tokens (NFTs) can have the character of collectibles. Graphic designers at larvalabs.com devised a program to generate an array of unique low-resolution stylized images of faces, each 24x24 pixels square. They called these NFT collectible Cryptopunks, which they describe as: "10,000 unique collectible characters with proof of ownership stored on the Ethereum blockchain. The project that inspired the modern CryptoArt movement."[25] At the time of writing, the cheapest reserve price for one of these images on their auction site was 80.95 ETH, the Ethereum currency, equivalent to $121,850.

Urban Non-fungibles

Cryptoart reinforces the concept of the artwork as a one-off, a unique item that can be gifted, passed on, inherited, bought, or sold. Art is normally *non-fungible*. Something is *fungible* if there's at least one correlate (or simulacrum) that can serve in its place as a substitute and that is useful in the same way. An automobile has many parts, and thanks to standardization and factory production, if one part wears out it can be replaced by another. Turning to critical art practices, apply fungibility to ideas about mass production and the replaceability of workers on the production line and you have the basis of Marxist critique.[26] That is commodification, a diminution of individuality, the denial of claims to uniqueness among humans and things—a denial of art according to some critics.

Fungibility is practical. Coins and banknotes are fungible. When I used to carry them, it didn't matter which £10 note I took from my wallet. It served the same practical function and hence value as any other in my possession.

Artworks preserve the idea that they are non-fungible. Such items are ostensibly unique one-offs, such as Rembrandt's *Return of the Prodigal Son*, the Cullinan Diamond, the Mask of Tutankhamun, a lock of Elvis Presley's hair—items of great value to some people, not least for their rarity or lack of portability. In the urban context, street furniture from a catalogue is fungible as it can be replaced readily. An entire building is less amenable to replacement. Real estate is non-fungible. It is easiest to think of a plot of land. It only exists in one place and at one time. Once someone takes possession of it then it cannot be possessed by another buyer. Of course, in the case of buildings, there are shared ownerships, variants, copies, and fakes but as with art, uniqueness is a major part of the value claim of urban real estate.

Technologies of reproduction distort the fungibility claim. We have 2D and 3D prints, photographs, scans, and other means of reproducing a thing, or an image of a thing. In the early days of online image sharing, I would read about the challenges posed by the new media. Some asked: Are we losing the concept of an original? In early reflections on the challenges of visual proof in the age of digital media in 2001, William J. Mitchell wrote, "Image files therefore leave no trail, and it is often impossible to establish the provenance of a digital image."[27] Now we have steganography, encryption, watermarks, and the blockchain. In spite of the proliferation of images, digital 3D models, 3D printing, and other means of digital mass production, no one seems really to have abandoned the idea of an original. Though originality may be harder to verify, digital media amplify attempts to discover and assert it. Creators want to share and disseminate, but they seek recompense for their labors, and are unhappy if someone else monetizes their production, or if they are denied a share, and their product loses value as its claim to rarity is diminished by profligate reproduction and sharing, especially online.[28]

NFTs offer the potential to secure intellectual property rights during the sale or distribution of BIM library elements (bespoke furniture, sculptures, doors, details). These are digital urban objects before they are manufactured as physical objects. Any document or data source could be so certified via NFTs. According to urban planning scholar Justin Hollander, "Blockchain technology and the use of NFTs to demonstrate ownership can be used to better track and manage property ownership and organize taxation."[29] NFTs are within the arena of smart contracts discussed in chapter 8. As with smart contracts, the field is open for research and exploration.

Steganography as a branch of cryptography operationalizes certain aspects of hidden layering. I think of steganography as an allegory for city form and function. It is also a potential aspect of urban screen culture to be exploited, detected, or resisted. Concealing images sounds suspicious in light of current social concerns about exploitative imagery. Working with blockchain technology, steganography, and NFT methods has the potential to certify, protect, and secure non-fungible urban assets. I will continue the general theme of authentication in the next chapter by elaborating on the instrument of the *hash*.

10 Hashing the City

A signature is an identifying mark. The cryptographic city, modern and ancient, is populated with signatures. I have mentioned my stone apartment built in the 1830s with the v-shaped mark chiseled into one of the time-worn stone steps leading to the front door. Around the corner is a garage door that the owner repaints periodically to obscure the graffiti that it seems to attract. Half a mile away there are rows of workers' homes that bear crests featuring the tools used by the tradespeople who once lived there. In Scottish towns you occasionally see markings in the stone lintels above doorways that indicate the date of the wedding of the original occupants. The shopping streets nearby are embellished with brand logos. In each case it's as if someone or some organization has left a mark in these places to assert ownership.

We sign off on a home delivery by leaving a mark on a piece of paper or touch pad or tap a button to show we have taken possession of some goods. Signatures authenticate our role in transactions that circulate by digital means and in digital formats in the cryptographic city. Non-fungible tokens (NFTs) as certificates of authenticity (COAs) expand on the concept of a signature to substantiate transactions. Signatures are a means of integrity checking and fall within the purview of cryptography.

A signature is a broad concept. It is common in architecture to speak of signature buildings, which serve either as markers of the company that owns or occupies them, or of the architect, developer, or benefactor responsible for their existence. I want to demonstrate in this chapter that in so far as our cities are permeated by digital transactions, they are imbued with signatures operationalized as hashes, passwords, and encryption keys.

Urban Profiles

In 1997, the Hubble Space Telescope's Space Telescope Imaging Spectrograph (STIS) picked up a signal of a black hole.[1] Instead of the usual vertical straight scan line, the STIS showed an S shape. It is common in astronomy and signal processing to call such a blip in a signal shape a "signature." Such signatures point to the presence of something. But you see the thing only indirectly, that is, you get to know that the referent exists, even though you can't see it simply by looking for it, or at it—even if you could. In this case the invisible black hole was such a referent. Viruses and other microscopic entities produce a genomic signature, or perhaps several, depending on the means of detection and measurement.

A signature can be physical or behavioral. Signatures have a broad range of uses and contexts of use. Some scholars have applied the term *signature* to large-scale human-made entities, such as cities, societies, or political movements. Many businesses, buildings and cities claim the attribution of signature synonymous with *landmark*, *showcase*, and *flagship*. That is to assert a signature object, like a building, as *significant*, as if it stands in for the larger whole, the enterprise, the collection, the style, or a nation. Some scholars even propose that places can have *affective* signatures. For example, an article on "emotional signatures" asserts: "Political ideology thus has a discrete emotional signature, one favoring anxiety among conservatives and anger among liberals."[2] Do cities have signatures? By some readings, "melancholy" is the emotional signature of cities like Porto in Portugal[3] and other cities bear affective signatures delivered in marketing, song, consumer comments on TripAdvisor, and Instagram tags, such as "delirious" New York, "friendly" Dublin, and "chilled" Amsterdam.

Digital technologies have expanded the opportunities to identify urban signatures. For example, scholars have detected the flow of people moving around a city derived from traces left across a map of the city by home delivery or taxi rides, or concentrations of images uploaded to photo-sharing websites.[4] Researchers have used these diagrams as indices of the city's cultural and social structure. Degen and Rose identify swirling linear graphics as a feature in marketing material for the smart city: "Digital data is made visible as geometric patterns of blue glowing light traveling effortlessly between various digital devices in the urban environment."[5]

Terms such as *footprint*, *fingerprint*, and *profile* also come to mind as synonyms for *signature* in an urban context. By this reading there are many signatures inscribed across the cryptographic city. In what follows I will focus on digital signatures and their role.

Scale and Compress

As indicated in chapter 9, drawings, photographs, image manipulation, and the application of graphic encryption bring us closer to the spatial aspects of the cryptographic city. City planners and developers identify densities and concentrations of populations and built structures, as well as the proportions of open to built space. In these and other respects, I think that aspects of urban space management are analogous to image manipulation. That helps me justify this foray into digital graphics as a facet of the cryptographic city.

If a signature is a surrogate for something larger, then a thumbnail image of a picture provides some of the functionality of a signature. Thumbnail images are simply reduced versions of a much larger file.[6] Such image reduction shows the functionality of displaying something large, like a picture that takes up most of the screen and reducing it to something small and more portable, such as a thumbnail image. Size reduction is among a family of techniques for turning an image into a signature.

Image compression, also known as *down sampling*, provides a similar function while retaining something close to the size and appearance of the original image. In that case compression algorithms down sample a full-color, high-definition image to a smaller file that is close in appearance to the original in terms of size and color range. The usual role of compression is to create a version of an image file that is as visually close to the original as possible but of a smaller file size.

There are several methods of image compression.[7] Lossless data compression produces a smaller file, but no information is lost, and display programs can reconstruct and display the image as if uncompressed. "Lossy" compression, such as JPEG image compression, adopts more cunning methods of file size reduction, such as reducing the palette of pixel color values where there are a lot of adjacent pixels with high contrast. The eye is less sensitive to color variation across a busy part of the picture compared with large areas of subtle color variation. We are more likely to notice dramatic contrasts than we are the subtle differences within regions of high contrast. A row of pixel values in a grayscale photograph will typically show a range of intensities. If you imagine these data points joined together by a smooth curve then that looks a bit like a fragment of an oscillating frequency curve, similar to a sound wave. There is an algorithm for breaking down complex curves into several regular cosine curves of different frequencies. The curve of the original row of pixels can be approximated as a series of numbers indicating the intensity of the contribution from several such regular cosine curves as sets of coefficients. Saving only the coefficients for the lower-frequency cosine curves reduces the file size. This cosine reduction method breaks the image into 8×8 pixel squares and applies the algorithm to each square in turn.

What I want to show here is that this compression method strips away a substantial part of the image information, thereby reducing the file size. Once it is lost, there is no way to bring that information back from the compressed file. The loss is more noticeable if the person viewing the image wants to zoom into the picture, enlarge it, or apply various filters as in a picture-editing program, or recover LSB steganographic image content as described in chapter 9, which will mostly be obliterated by JPEG compression.

Scaling down the size of an image involves a simpler method than JPEG compression. It simply averages the color value of every square group of pixels (e.g., 8×8) to a single pixel. The averaging process loses information. It is not possible to reconstitute the original information from a set of averages, though it is usually recognizable by a human being as a surrogate for the original image. A thumbnail is such a reduced-size version of an image.

These techniques introduce the idea that you can reduce the size of an image by scaling, image compression, and other manipulations. The process leads eventually to something small and unrecognizable by the human eye as derived from the original referent. But this much-reduced image serves as a digital signature, a pointer to the original, a means of authentication and a file that is more portable than its referent. Importantly, the creation, storage, sorting, manipulation, and matching of this signature can be automated.

Picture Hashing

The kind of signature I have just described serves some of the functions of a *hash*. An image hash is a reduced version of an image file and serves many of the functions of a signature with additional benefits, such as matching, indexing, lookup, and search. The hash image provides an almost unique identifier for an image in a lookup table or index, and the hash may preserve some of the properties of the source image, such as its average color, and the general distribution of colors across the image. Image indexes can deploy searchable lists of hashed images. Matching hashed images narrows the search process. The use of the hash from the point of view of graphical images is relevant to visual images, maps, remote sensing scans, photographs, screen displays, and other visual representations on screens on smartphones, laptops, and urban displays.

I introduced the hash idea in chapter 4 as it accords with concepts of urban collage, and in chapter 7 as I was explaining the authentication of blocks of transactions on the blockchain.[8] There I was referring to a hash as a sequence of characters of fixed length that is derived from the contents of a file, whether text, image, or other medium. Though blockchains can reference images, they deal ostensibly in strings of alphanumeric characters, of text documents. Another important function of a signature is that of authentication. I referred to the challenges of authenticating crypto art in chapter 9 via certificates.[9] In the rest of this chapter I will elaborate on the hash idea, already introduced in relation to the blockchain in chapter 7.

Hashing on the Blockchain

To hash a file does not encrypt it, as there is no way to reverse the process to recover a file from its hash. Hashing assists in validating data. However, the hash function is one of a range of tools that cryptographic security systems use, not least to protect passwords, verify the integrity of data transfer, and to construct the blockchain.

There are different hash standards, and well-known algorithms that generate hashes from text or other data.[10] As an example, the SHA256 standard algorithm will always convert "abc" to a certain bit string (1s and 0s) 256 binary bits long generally printed as a hexadecimal number of 64 characters:

ba7816bf8f01cfea414140de5dae2223b00361a396177a9cb410ff61f20015ad

The phrase "cryptographic city" converts to the equally long and arbitrary looking ab4588fce73f79a6ef7bf717872ae010203b8f26e7ba9df71dd14028a09be5f0

To illustrate the generality of the method, here is a section from a non-confidential government document outlining a graffiti strategy for Edinburgh.[11] Though the content is interesting from the point of view of signatures in the city, I am only interested here in this passage as a character string:

> Current procedures and guidelines have been reviewed and best practice identified to ensure that a balanced approach is taken. Robust policy(s) and procedure(s) on Graffiti Management are key components of the future strategy, aiming to reduce instances of "tagging," while still providing space for the more creative elements and potential benefits of graffiti, street art, and murals for local communities. This will also ensure that the city's residents and stakeholders are clear on the approach being taken by the local authority.

Under the SHA256 hashing algorithm that passage translates to the random-looking string

1d6d8d911d3fa0b03af4791c99ea3f3c56f6ecf0d4cc7f201d9cf3783948cce2

The 64-character string output serves as a fixed-size unique condensed representation of the original text about graffiti strategy. Take out the first full stop in the passage and the hash algorithm produced a different string output:

8a5d4771399383e4535bec29432aab2e8d5ac63dafad83844b3fb0b01eaf5136

Neither human nor algorithm can detect any pattern in those strings that indicates the proximity to their original source strings. The conversion is not a method for compressing or encrypting data. There is no algorithm to unpack the 64-character string to recover the original text. The output string is for machine consumption rather than for human beings to read, check, and ponder. The output string is the *hash*, and the SHA256 algorithm is one of the most sophisticated and the least likely to generate "a collision," meaning, where two or more sets of different data have the same hash. The SHA256 algorithm is no secret, and there are many explanations of it in cryptography textbooks and online,[12] and knowing how the routine works to convert a string of text to a hash string provides no advantage to a would-be hacker.

What is the use of this (near) unique hash representation of a block of data? It is used for authentication. As well as transmitting the document, you transmit the hash—perhaps in a different channel. The receiver software then runs the data through the SHA256 algorithm, and if the receiver-generated hash does not match the hash sent with the data, then that alerts the receiver that the data has been tampered with en route. It won't tell you what was changed, or the magnitude, but it may send an alert, or notify the sender software to transmit the data again.

Here is a graphical example of the use of the hash algorithm. The CryptoPunks auction site sells CryptoPunk NFT files. As introduced in chapter 9 these are stylized 8-bit portrait images each sized 24 × 24 pixels. The auction site describes how they use hashing so customers can verify that once they have made a purchase they are in possession of an actual CryptoPunk file: "The actual images of the punks are too large to store on the blockchain, so we took a hash of the composite image of all the punks and embedded it into the contract. You can verify that the punks being managed by the Ethereum contract are the True Official Genuine CryptoPunks by calculating an SHA256 hash on the cryptopunks image and comparing it to the hash stored in the contract."[13]

You don't need to purchase a CryptoPunk NFT file to test that the file you have downloaded is exactly the same as in the Ethereum contract. The Ethereum contract cites the hash of the composite graphic file displaying the composite of ten thousand CryptoPunk images as

ac39af4793119ee46bbff351d8cb6b5f23da60222126add4268e261199a2921b

The composite CryptoPunk file is a picture, but it is just data as far as the SHA256 algorithm is concerned, a file of 0s and 1s. Sure enough, the CryptoPunk graphic file from the LarvaLabs website returns the identical hash string when I run it through an independent SHA256 calculator.[14]

A hash can also be used to transmit a password over the Internet. My online banking service has my password in a database saved as a hash. So, if I had a password such as "appLeTr33" and wanted to log on to my digital bank website then the transmission software would convert that password to the hash

80dfcee4ead6ef8cd1ba4edfc782d5c4b4505affc223d1345bcce51c9818fc61

and send the hash through the Internet. The server software receives the hash and looks that up to access my account. If the hash falls into the wrong hands, then there is no way to reverse engineer the hash string to the original password. So that's one of several measures required to keep passwords safe in transit.

Hashing is also used for indexing whole files via so-called "content addressing," the unique hash as a signature derived from running an entire file through the hashing algorithm. IPFS, the so-called "Interplanetary File System," is a peer-to-peer file storage and sharing platform. The system breaks computer files into segments (256 kB "chunks") and distributes them to different servers on the network. The segments are indexed via their hash so they can be reassembled and accessed via their hash strings, which can in turn be secured on a blockchain.[15]

So, I have introduced the role of the hash in the blockchain and alluded to other processes pertinent to the blockchain. The rest of this chapter provides a deeper dive into blockchain processes for readers who share my enthusiasm for cryptographic methods.

Hash Chains

Table 10.1 shows a page from a spreadsheet of transactions. It is technically a Merkle chain.[16] I have added a column to show the hash of the data in each row, added to the hash of the previous row. You can create a hash of a hash, a process that can be continued indefinitely. If someone changes the content of one of the rows in the ledger then that changes the hash values for the transactions that follow, as shown in the second part of table 10.1 (doctored ledger). With access to the hash algorithm, someone who wanted to change a transaction in the ledger could change the hash value of that transaction and those that follow. But if the ledger has already been shared with others on a network and verified, then the discrepancy will be obvious, and rejected as invalid by auditing algorithms in the network.

In fact, this method of verification via hashing works on whole pages of transactions as well as for individual transactions. One of the best descriptions of this method of verification that I have found comes from a post by blogger Antony Lewis.[17] He explains that a traditional paper ledger consists of pages of transactions. Instead of page numbers, at the top of each page (header) print a hash of the previous page. At the bottom of the page print a hash of that page (including its header). Lewis calls the hash a "unique fingerprint" of the page: the page is a block, the whole set of pages linked in this way is a chain.

Table 10.1

Securing data on a Merkel chain

Original ledger of transactions

Date	Description	Value	Previous hash	Combined record	New hash
12-Jun-22	Accommodation	−$210.00	0000000	12-Jun-22 Accommodation −$210.00 0000000	63896d1c
14-Jun-22	Bike hire	−$67.00	63896d1c	14-Jun-22 Bike hire −$67.00 63896d1c	a36aca28
17-Jun-22	Refund	$50.00	a36aca28	17-Jun-22 Refund $50.00 a36aca28	fe52f149
19-Jun-22	Home delivery	−$35.00	fe52f149	19-Jun-22 Home delivery −$35.00 fe52f149	561dc761
21-Jun-22	Expenses	$250.00	561dc761	21-Jun-22 Expenses $250.00 561dc761	07dc2d3d

Doctored ledger

Date	Description	Value	Previous hash	Combined record	New hash
12-Jun-22	Accommodation	−$210.00	0000000	12-Jun-22 Accommodation −$210.00 0000000	63896d1c
14-Jun-22	Bike hire	−$67.00	63896d1c	14-Jun-22 Bike hire −$67.00 63896d1c	a36aca28
17-Jun-22	Refund	**$100.00**	a36aca28	17-Jun-22 Refund $100.00 a36aca28	**2a619b4a**
19-Jun-22	Home delivery	−$35.00	2a619b4a	19-Jun-22 Home delivery −$35.00 2a619b4a	**2a2e40eb**
21-Jun-22	Expenses	$250.00	2a2e40eb	21-Jun-22 Expenses $250.00 2a2e40eb	**60fe8744**

The first section shows a page from a spreadsheet of transactions with chained hash codes. The doctored section shows a spreadsheet with the "Refund" transaction altered. That change alters the subsequent hash codes in the chain. In a shared ledger system, the change would be rejected. Note that for simple illustration the hash code used here (CRC-32) is much shorter than the SHA250 standard.

Internal consistency is assured if the hash strings or "unique finger-prints" are consistent with the data and they chain together effectively. In order for someone to adjust the ledger they would have to adjust all of the chained hash strings that follow the adjusted transaction. The algorithms of the blockchain would soon detect the inconsistency in the final state of the blockchain and reject the rogue transaction.

The term *immutable* is used commonly to describe this state of a data set.[18] Each page of transactions is embedded in the transactions that follow. So, you end up with a very large unalterable database. As discussed in the context of the blockchain in chapter 7 it is a bit like sedimentary rock formations where the newest strata depend for their integrity on older formations below, which in turn require more effort to penetrate, or alter. As we have seen, this "immutability" is a key feature of distributed digital ledgers as used in cryptocurrencies, where the ledger is duplicated and shared across a large number of users, and there is no centralized keeper of the data or auditing authority such as a bank. Hashing secures the integrity of the blockchain.

Hash Puzzles

To reverse the process of bedding transactions into a blockchain, and to unpack the original string from a hash would be a mammoth computa-tional task, requiring software to generate every combination of characters, of different lengths, and calculate the SHA256 hash of each combination to see if they match the target hash string. Even if this brute force search yields results for hackers, there may be several strings that produce the same hash. So, they cannot be sure they have found the right one. The SHA256 algo-rithm is designed to make this reverse engineering virtually impossible. Unlike various computational puzzle-solving tasks, the algorithm would be unable to detect that it is getting closer to a solution. It might as well iterate every combination of characters for the input at random.

The proof of work (PoW) verification procedure recruits a simpler hash puzzle: find an input string that generates a hash with certain char-acteristics. For example, generate a hash string that starts with a zero. I experimented with an arbitrary string of characters "301253" by chang-ing, adding, and deleting characters, fed each string through the SHA256 algorithm, and eventually and randomly came up with "301252" that generates this hash:

0135ce9c1094a7b56219acca34ba9e3bb68bd883cf842577c1e8c107346af173

That's a hash string starting with "0." I solved the puzzle manually by pasting strings into the online SHA256 calculator.[19] But the process could be automated. Imagine this is a contest in an online multiuser game. The first person to "solve" the puzzle of producing a hash string that begins with 0 gets a game reward—some points. Once you guessed an input string that produces a hash string starting with zero, you would send around your input string (301252), and the rest of the players could easily verify that you came up with an answer. They would run the string you distributed through their own copy of the SHA256 algorithm to see if it did indeed generate a hash starting with zero.

In the preceding case the challenge was too easy, computationally. You could set a harder challenge: what input string will generate a hash string that starts "01010," "ababab," "0000," or some other sequence? Aiming for a hash that starts with a row of zeros is as good as any target, as it's easy to verify by eye if needed in a demonstration, and the degree of difficulty will be obvious from the number of zeros required. One leading zero presents a trivial challenge; four zeros automatically increase the degree of difficulty and the number of iterations required considerably. If the challenge appears to be too easy, solved too quickly, then an algorithm shared across the network will increase the degree of difficulty of the challenge by increasing the number of zeros required. It seems that cryptocurrency systems that use this method expect solving the challenge to take about ten minutes of dedicated CPU time, which is a lot of brute force computation. The competing computers are the validating nodes, the miners on the network that supports the blockchain. As faster CPUs enter the network, an algorithm will increase the degree of difficulty to match the capability of the competing validating CPUs. The degree of difficulty for the Bitcoin system is at the time of writing set at a hash with nineteen leading zeros![20]

As someone invested in architecture, design, and gaming, I and many others find it difficult to warm to this extravagant use of processor time in an invisible contest that yields no nuance, color, spectacle, or entertainment. In fact, during lockdown I tried to purchase a PS5 gaming console without success. Though it has no need for graphic functionality, crypto-currency mining favors the use of fast graphics processing units (GPU processors). GPUs were in high demand for cryptocurrency mining, resulting in a serious undersupply of gaming consoles.[21]

Proof of Work

What I have described involves securing a page of transactions via a trivial but time-consuming computational puzzle. I have demonstrated how a page of transactions can also be hashed. A page of data in the blockchain includes in its header a hash of the previous page followed by a list of transactions.

To summarize, this puzzle-solving is used to verify and embed a set of transactions. Cryptocurrencies, such as Bitcoin store-linked pages (blocks) of transactions and distribute these to all the validating nodes connected together on a vast peer-to-peer network. Any node on the network, usually a self-appointed subset with adequate computing power, can use the solution to this puzzle to bed down a set of transactions. The node must collect all the transactions coursing through a network since the last page was formed, check that they are consistent and legal, and put them into a virtual ledger page (a *block*) in a standard format. Many of the independent nodes will be doing this at the same time, incentivized by the potential financial reward. So, it is a high stakes contest.

As soon as one of these nodes generates a solution it broadcasts the result to all the other nodes, which quickly verify that the answer is correct. It is easier to test that the solution is correct than to discover it. The block of the winning node then gets added to the set of all approved ledger pages, which is in turn distributed around the network as the approved set of transactions making up the correct and current state of the ledger. That is the blockchain.

As already indicated, the work required to bed down the pages in the ledger (i.e., verify, approve, and seal the blocks in the blockchain) is the proof of work. Were any potential hacker to try and overwrite any transactions or otherwise doctor the ledger, from the most recent page going back many other pages, then they would have to adjust the whole sequence of hashes in all the blocks. According to the seminal article on the technique by Nakomoto, "To modify a past block, an attacker would have to redo the proof-of-work of the block and all blocks after it and then catch up with and surpass the work of the honest nodes. . . . The probability of a slower attacker catching up diminishes exponentially as subsequent blocks are added."[22]

There are other sophistications to the process, and I've simplified some steps and concepts, but I think the explanation provides at least an insight into the purpose of setting arbitrary puzzles to keep transactions safe from tampering, a necessity where ledgers are open, duplicated, and distributed across networks, breaking the need for a single "trusted" custodian of the database, such as a bank. The deliberate and extravagant "waste" of CPU time is put to a purpose. It is part of the cost of security in a distributed peer-to-peer data system. I mentioned an alternative to PoW in chapter 7, namely proof of stake (PoS).[23] Both systems make use of hash algorithms to embed blocks of transactions in the blockchain, though the PoS draws on other computational methods that require less CPU time in validating transactions.

In this chapter I have demonstrated that the idea of the signature is at home in the smart city, especially in its cryptographic instantiations as hash strings. It also permeates urban digital infrastructures and security and validation processes. To the extent that cryptocurrency is prominent in the cryptographic city, the processes I have described assume significance as city infrastructural elements. They also parallel certain urban processes that run deep in the economy of the city: competition, networks, ledgers, hidden processes, opacity, in the communicative functions of the cryptographic city.

IV Cryptography at the Limits

11 Cyberattacks

One way to conceal information is to position it within a confusion of signals (that is, to obfuscate). Hash strings obscure the content of the files that they reference. Steganography, hiding one image inside another, obscures a secret image within the complexity of the carrier image that conceals it. A maze will obscure access to the center, the tower, the keep, or the gate to the city via the confusion of twists, turns, and junctions. Spatial confusion is one of the means of defense and protection in cities as in nature.

Confusing the perceptions of predators and prey is a basic device in species selection. Any single zebra will blend in with the herd when they stand together. It is harder to tell where one zebra ends and the next one starts. As they approach the herd, the visual field of a lion or hyena is assaulted with vertiginous to-and-fro movement—which is somewhat stroboscopic. Any predator will be confused, just enough, to give the herd time to take evasive action. Such momentary obfuscation buys time.[1] Obfuscation is a wily tactic exercised by both predator and prey. It is also a feature of accidental and deliberate human tactics of attack and evasion (figure 11.1). One of the strongest means of attacking an adversary, target, or victim is to throw them into confusion.

Markets, fairgrounds, malls, and busy supermarkets lure shoppers with sounds, colors, and spatial confusion to provide an entertaining melee. They also disorient shoppers as prey for sellers and advertisers. On the other hand, multiplication obscures the prey. A would-be thief will find it harder to identify a particular victim in a high-density neighborhood made up of hundreds of apartments than in a sparsely populated rural setting.

Factors other than protection contribute to spatial obfuscation in the urban context. The exteriors of buildings commonly conceal their functions.

Figure 11.1
Dazzle obfuscation. HMS *President* docked on the River Thames, London, as a recreational venue and painted to resemble tactics for camouflaging ships to confuse detection by adversaries. Photograph by Ron Ellis via Shutterstock.

A building designed with multiple readings in mind constitutes such a dissimulation. In the old part of Edinburgh there are recesses in external walls that are the same shape and size as windows. That is to preserve the regular patterns of openings across a wall, even where there's a chimney or fireplace on the inside that blocks the possibility of a window opening. Cities as sites of communication are populated by mimetic, faux, mock, and dissimulating architectural signs that occlude the legibility of the city. Obfuscation is also a prominent tactic for attack and defense in the online world, and hence in the cryptographic city.

Cryptography is a subspecies of obfuscation. After all, much encryption turns messages into apparent noise. Cryptography invokes an explosion of possible combinations designed to overwhelm the cognitive and computational capabilities of humans and computers to sift, sort, identify, and edit signals and messages.

A Darker World

By some philosophical readings, the world is a mass of confusion, and clarity is the exception, imposed by sophisticated cultural framings, and language.[2] To obfuscate is "to confuse, bewilder, or stupefy," according to the OED. But the first definition is "to darken," from the Latin *obfuscare*. From a phenomenological reading, our perception of the world can usefully be considered as a process of revealing and concealing. As the spotlight of human perception and insight scans across the world it reveals certain objects, attributes, and ideas, but others recede into shadows and complete darkness. The world is already a sea of zebra stripes. It is our lionlike acuity in marking out the exceptional, the independent, and the adventurous that constitutes the unusual moment of lucid perception.

Obfuscation is effective as a tactic to confound the mind of the rational citizen. An intriguing book by Finn Brunton and Helen Nissenbaum outlines various obfuscating tactics and their remedies and ethics: *Obfuscation: A User's Guide for Privacy and Protest*.[3] Here, obfuscation is a way of dealing with information. It is semiotic. It pertains to information that we want to conceal. In the zebra case, the information is the location of one vulnerable, singular individual that is potential prey. For Brunton and Nissenbaum, "Obfuscation is contingent, shaped by the problems we seek to address and the adversaries we hope to foil or delay, but it is characterized by a simple underlying circumstance: unable to refuse or deny observation, we create many plausible, ambiguous, and misleading signals within which the information we want to conceal can be lost."[4] Their text is a handbook to help readers avoid the inadvertent disclosure of personal online data, including one's location, and other information by which advertisers, government and foreign agencies, political operatives, and hackers can surveil and profile us. The authors provide several means of obfuscating data.

Jobsworth

Brunton and Nissenbaum show how you can obscure your actions and intentions by adhering to the letter of the law, to comply with formal requests, to do what you are told, and to deliver truthful information—and you can confound your antagonist by doing so. I think anyone who works in a large organization knows how compliance can obfuscate, particularly in relation to information: provide the information requested—even more information than is asked for; provide it unfiltered and unsorted; inundate your antagonist with "paperwork" and keep it coming; exploit the fact that the requester may not really know, or forgets, what they are looking for or why they requested it. After all, you were only asked for the information, not to be helpful beyond that.

Here is a variant of this obfuscatory tactic. Brunton and Nissenbaum point to the use of vague language in a promise. It looks as though you are saying what people want you to say, but you say it in such a way that it provides you with a loophole in case you cannot deliver on your promise: "I promise to negotiate a deal that will deliver the best outcome." Is the promise to negotiate a deal, to deliver the best outcome, or both, and the best outcome for whom? That's my example. They provide the example of a web service that asserts, "Certain information may be passively collected to connect use of this site with information about the use of other sites provided by third parties."[5] That looks to be showing responsibility and care for user privacy, in that the website is telling you something. But it is confusing. What we would like to read is: "We do not collect user data," but they probably do. Unfortunately, they don't say that they do not.[6]

I particularly like Brunton and Nissenbaum's advice on spreading culpability to protect the guilty. So, activists in a group wear the same clothes as each other, including face coverings. Then it is harder for witnesses to identify the individual who actually threw the eggs at the politician, smashed the plate-glass door, or spray painted the shop window. An obvious variant is for the individual activist to appear indistinguishable from the innocent crowd. In fact, certain offences are best committed where there are crowds. If you are going to do something against the law, then look the same as everyone else. The "I am Spartacus" tactic is a variant. Only one person is guilty, but every member of the group or the extended group of sympathizers confess to the crime. Outside of despotic Roman "justice," they cannot all technically be guilty, or punished.

The identification and arrest of rioters after the U.S. Capitol security was breached on January 6, 2021, highlights the extent to which the proliferation of online photo sharing, live streaming, surveillance cameras, police body cams, and press coverage subverted the aims of individuals who wanted to get lost in a crowd.

Brunton and Nissenbaum's book is ostensibly about how citizens and activists can evade and subvert how commercial interests and governments collect
data about them. But bad actors deploy the same tactics. In April 2019, the
U.S. Department of Justice released a report on Russian interference in the U.S.
presidential election of 2016, authored by a team led by Special Counsel Robert Mueller.[7] A witness in the Mueller investigation said that President Trump
"took every step that he could to try to obfuscate, to try to get people to lie,
tried to reward those people who refused to cooperate with a legitimate investigation, tried to punish and denigrate the people who were cooperative."[8] That
is how one witness summarized the Mueller Report. I searched for "obfuscation" in the report and it's not there, but "obstruction" features prominently.

Brunton and Nissenbaum's book was published before Trump was
elected, and before the concept of "Active Measures" gained currency among
the general population. These were the operations of the Russian Internet
Research Agency and of the GRU (General Staff of the Armed Forces of the
Russian Federation) that include tactics to confuse, obstruct, and obfuscate.
Brunton and Nissenbaum's book addresses different sides of such operations. On the one hand are powerful predators who want to exploit us and
make money or extract political advantage from our personal information.
On the other hand are those ordinary citizens and (apparently) good-faith
activists who want to confound the attempts by powerful corporations,
organizations, and states to monitor, surveil, profile, and coerce us.

Practitioners of cryptography want to both obscure secret messages and
obscure the fact that there are secret messages. Codebreakers and hackers
want to conceal that they have stolen or copied some data. They also want
to conceal secret code and stegomalware they have inserted into digital
assets and to cover their tracks, concealing that they were even there. Software platforms are difficult and confusing in any case with contributions
from multiple players and under the control of operating systems, subrou-
tine libraries, interface protocols, and networks that are in turn developed
and maintained by a circle of independent suppliers. Software is an ideal
medium in which to employ obfuscation and espionage.

Automated Obfuscation

Brunton and Nissenbaum describe *chaff*, a technique to confound radar
by scattering scraps of foil-backed paper into the sky. Chaff is a simple,

low-tech method to scramble radar signals to make it more difficult for an adversary to detect the location of your attack plane as it approaches its target. The point is not to hide the aircraft, but to make it appear there are more planes than there really are. That is a trick of obfuscation: not to hide, but to multiply, and thereby overwhelm the system of detection. Technologies are good at multiplying and repeating. Eventually the recipient of the obfuscation tactic gets the data they seek, but it takes time, even for detection technologies. As for a herd of zebra, obfuscation serves to delay rather than prevent detection.

There is a zebra versus lion, prey versus predator contest in play, as each vies against the other with ever more sophisticated means of evasion and capture. It is a basic contest played out over various scales of technological sophistication. Brunton and Nissenbaum provide many examples of the ploy by both good and bad actors, predators and prey, the powerful and the less powerful, the corporatized and the independent, the seller and the consumer, the saboteur and the victim.

The deployment of Twitter bots provides an obvious example of obfuscation by digital means. Fake twitter accounts generate new tweets. They retweet the tweets of others, select from catalogues of standard tweets, generate likes, and generate new fake accounts to compound and confuse social media messaging. Such tactics attempt to skew people's impressions toward a particular point of view, to exaggerate the apparent support for one opinion or person, or simply to confuse the audience. The Mueller Report made clear that bad actors seek both to conceal (obfuscate) their own operations and to obscure the public discussion by generating fake support for conflicting opinions. The adversary seeks to divide a population and sow chaos.

When the U.S. Department of Justice first made the Mueller Report available to journalists and the public it was released as an image file. Every page was a digital optical scan, but you couldn't search that for words or phrases. Whether deliberate or not, the format served initially as a means of obfuscation. Soon after its release I tried to run the file through Adobe's optical character recognition (OCR) function, but the file was too large for my version of the OCR reader at the time. Eventually, a week or so later, a searchable version of the report appeared online.[9] The use of technological impediments served as a means of further delay. As with encryption and decryption methods, changes in software, standards, and systems impact tactics for obfuscation and breaking through the noise.

Cyber Espionage

I have already alluded to how bad actors and agents of espionage might deploy obfuscation. The respectable-sounding Internet Research Agency (IRA) is a media organization that was started by the Russian government in 2013, initially to exert influence over Ukrainian and Russian citizens.[10] It has since rebranded and bears the name of its location in the landmark business center Lakhta in St. Petersburg. The 2021 U.S. NIC (National Intelligence Council) report into interference in the 2020 U.S. federal elections describes the organization as "The Kremlin-linked influence organization Project Lakhta and its Lakhta Internet Research (LIR) troll farm."[11] Before the 2016 U.S. presidential election the Russian IRA directed its operations to influence online political discussions in the United States, with further influence in other countries, though that hasn't drawn as much attention in the press. Reports released in 2018 referred to so-called "organic" online activity that included innocent content supplied by consumers as tweets, Facebook posts, blogs, and YouTube clips, as well as comments, reposts, links, likes, followings, and subscriptions.[12] So "organic" online activity embraces noncommercial consumer activity, in contrast to advertising activity paid for by sponsors that usually appears conspicuously as banner ads, pop-ups, inserts, and branded clips in news feeds and videos.

Advertisers seek to persuade and to reinforce a brand. Political parties and other interests deploy a range of tools of persuasion and for propaganda. They may also seed organic posts by legitimate audiences and customers. Organic customer engagement strategies enhance the reach of the brand with only modest financial outlay.

The reports I referred to earlier deploy the term *organic reach* to describe the processes by which advertisers or any organizations exert influence by pretending to be social media consumer sources. They deliver false consumer-led online activity such as recommendations, or just drop propaganda or endorsements into tweets, posts, videos, and online conversations. Covert and fake organic tactics of the kind deployed by the Russian IRA included posts, feeds, comments, and other content, including likes, each delivered from false or misattributed accounts, organizations, communities, and individuals. A malign organization can introduce paid ads that deliver misinformation and that claim to be from a source other than those running the influence operation. Bots, algorithms, and pools of human operatives pretending to be legitimate social media users simulate organic activity and seed further reach.

So-called "voter suppression" was one of the tactics of Russia's Internet Research Agency to discourage people in the United States from voting. The tactic was directed at people on social media within a demographic that can be identified readily and is usually inclined to vote in a particular direction. For example, the average African American voter is or was assumed likely to vote Democrat in certain US states. To suppress that vote, the malign agent presents to members of that group "tweets designed to create confusion about voting rules," according to the report *Tactics and Tropes of the Internet Research Agency*.[13] Voters might be led to believe falsely that they can get someone else to vote on their behalf, or that they can deliver their vote online. The Russian IRA might also encourage that demographic to vote for a third-party candidate or an independent minority candidate, thereby diluting the support for a main party candidate. The third method here is to persuade members of that group not to vote as their vote will not make a difference anyway. That is one example of how such an externally generated influence campaign can work. Combined with sophisticated monitoring of social media users, targeted organic reach, and hacking into private records, we have the makings of cyber subversion, if not cyber warfare. Obfuscation serves as a means of espionage and it is among a series of measures by which foreign agents confound local, urban sociability and security in the cryptographic city.

Active Measures

Political commentators identify how Russia's IRA deployed long-standing Cold War tactics to induce foreign targets to disclose secret information and to spy on operations in their own countries.[14] Active measures include well-resourced propaganda and influence campaigns operating across media channels including social media. Some techniques of "active measures" focus on *kompromat*, a variant of компромисс, the Russian word for "compromise," various means of accessing and activating embarrassing and damaging information about individuals and groups. Not all active measures are effective all the time, but are cheap to implement and are much less costly than planes and artillery. In any case, if a tactic does not succeed, the main return is confusion among its adversaries. If such "active measures" are delivered by one government against another, and the malign government is an autocracy, then its state-controlled media cushions any bad publicity that comes its way. That malign government can also keep its operations hidden, as it infiltrates the global Internet and social media with propaganda and misdirection.

An illuminating article by Pawel Surowiec shows why active measures and kompromat are difficult to counter: "There is a risk that countering *Kompromat* inspired propaganda head on will lead to the proliferation of the very information one is trying to counter in cyberspace."[15] Cautious commentators will state that they don't want to restate the lies out loud as that increases their circulation. The article by Surowiec was published in Autumn 2017, the early months of the forty-fifth presidency, and events moved quickly: "*Kompromat* is a flexible and powerful concept. It enables denial (rarely apologia) of any wrongdoing when uncovered. Additionally, it often reveals falsehoods and lies about political or business opponents along with truthful negative information, blending accuracies and misinformation, thus allowing it to damage its targets in a highly sophisticated manner."[16] Kompromat obfuscates.

For those attracted to conspiracy theories, it is indeed peculiar that made-up stories about a "deep state" and the free press as "enemy of the people" had greater circulation than the more plausible, fascinating, and "real" narratives that identified the culpability of kompromat campaigns. But deflection from its own operations is also part of the kompromat play— the circulation of distracting and unsettling kompromat-inspired false accusations and conspiracy theories.[17]

Zero-Day Attacks

The repertoire of malevolent active measures by states and criminal opportunists includes those that target cities. Cities and computer systems are complicated, with legacy components, new and old bits of code, tangled communication routes, and gateways and networks that make them vulnerable to bad actors and resourceful hackers. Overlay such digital vulnerabilities onto the arcane and labyrinthine structures of cities and infrastructures that have grown piecemeal over time. That gives you the hackable city, a city that is victim of its own obscure legacies and substructures.

Some espionage tactics exploit complexities and vulnerabilities in computer systems. The concept of "zero-day vulnerabilities" emerged in the 2020s. It is usual for computer code to have "bugs," no matter the level of quality control and how well it is written. Any computer interaction, utility, or function depends on components supplied from a range of sources. Operating systems provide the environments in which programs function. Programmers draw on shortcuts to various functions such as libraries of subroutines. A weakness or incompatibility among any of these components introduces errors. From the point of view of an end user on a word processor or data entry package, the software may simply "crash," or a function may fail.

Thanks to online monitoring and harvesting of user feedback, software suppliers develop and distribute regular "patches" that fix or replace malfunctioning software components. That works as a means of maintaining software quality as long as the supplier finds out about the bug before it causes serious damage, including reputational damage to the supplier's product line.

Some malfunctions have implications that extend beyond inconvenience for a single end user. These are vulnerabilities that provide portals into software and systems for the spread of malware that cripples the digital functions not just for the individual using the software, but also for networks of users, organizations, countries, nations, and even global systems. According to cybersecurity lore, in the digital world it is safest to assume there are bad actors, operating alone or in groups dispersed across networks, who operate as rogue employees of government, or who belong to both benign and rogue states. All have the capacity to detect and exploit vulnerabilities in software.

We can also assume that there are many points of entry to any networked system. Software, such as the Microsoft operating system, is widely distributed. Not all individuals and organizations run the latest versions of the software or institute all patches and fixes when available. Vulnerabilities concentrate at the least secure nodes in such networks.

How do rogue agents identify these vulnerabilities? There are many methods. One obvious route for a hacker is to steal the source code somehow and analyze it for potential failure in the event of bad input data that it cannot trap and that causes memory overflows, for example. But the most common method of attack is "fuzzing," "a brute force approach in which the attacker provides overly large or otherwise unanticipated inputs to a program and then monitors the response," according to an article on such vulnerabilities.[18]

Once the hackers know what makes the software system fail then they can hold the software to ransom for the users who depend on it, either by threatening to disable the software, or encrypting its data so that the company or user being targeted needs to pay for the key to restore it. This was the tactic in the case of the malware breach of the UK's National Health Service in 2017. The so-called *WannaCry* ransomware attack exploited vulnerabilities in the Microsoft operating system. The U.S. National Security Agency (NSA) had already detected the vulnerability and kept that knowledge secret in case they could exploit it as part of their own cybersecurity defense and attack weaponry.

Hackers had also come across the vulnerability, perhaps through a leak from the NSA or by some other means, and exploited it by disabling crucial systems and demanding payment in exchange for restoring the data. Software users were outraged by the hackers of course, but had the NSA reported the vulnerability to Microsoft, which would then have patched it, this would have spared the NHS and other organizations the ransomware scam. The NSA had nicknamed the Microsoft vulnerability "Eternal-Blue." Information scientist Stephen Wicker explains the problem: "Having learned of (or discovered) EternalBlue, the . . . perpetrators used the vulnerability to put target machines in the desired vulnerable state, and then issued a 'request data' command that caused an encrypted viral payload to be loaded onto the target machines. The payload included ransomware as well as software that searched for other machines that had the same vulnerability. The ransomware rapidly propagated across the Internet, infecting machines that shared the EternalBlue vulnerability."[19]

The offense does not stop with the data hack. Knowledge of vulnerabilities is a marketable commodity. Users, systems operators, or hackers may happen upon these vulnerabilities by serendipity, or may actively seek them out. Once detected, we might think it appropriate to report any vulnerabilities to the software supplier, which would then write code to obviate the problem and distribute that as patches or upgrades. That may take a few days or weeks, during which time the security of the software and the systems that depend on it are compromised. There are *zero days* to fix the problem, hence the naming of the zero-day vulnerability.

Cybersecurity critic Nicole Perlroth explains the phenomenon of zero-days incursions into the functioning of major software systems: "They are a cloak of invisibility, and for spies and cybercriminals, the more invisible you can make yourself, the more power you will have. At the most basic level a zero-day is a software or hardware flaw for which there is no existing patch."[20] Cybersecurity advocates might argue that the security of citizens is served well by government agencies able to exploit vulnerabilities in the software and systems of hostile foreign actors. The information has value to the NSA or other state instruments that can exploit the vulnerabilities in disabling the systems of their adversaries.

As well as the software developers and suppliers, third parties are interested in knowing about the vulnerabilities. An online article by Matt Suiche in 2016 revealed that a group of hackers known as the *Shadow Brokers* detected the EternalBlue vulnerability and offered the information for online auction.[21] They might have sold that information to the NSA or perhaps the information was sourced from the NSA in the first place.

The infrastructures of entire cities have been compromised through ransomware exploits. Government security agencies have to deal with the question of whether they should disclose their knowledge of any vulnerabilities to the vendor, or keep it to themselves to aid their own covert operations.

Bulk Surveillance

Nation states have departments such as the NSA and the UK National Cyber Security Centre that commit to keeping national communications networks and hence city infrastructures secure. They invariably deploy cryptography in both defensive and offensive measures. In 2013 one of the NSA's contract employees, Edward Snowden, disclosed NSA covert operations via the press in a series of consequential revelations. He was pursued by law enforcement and eventually moved to self-imposed exile in Russia. He explained his reasoning and psychological state in an autobiography, as well as several documentaries and press interviews.[22]

To provide an urban context, as a fugitive Snowden declared that when crossing a busy road, he instinctively looked away from oncoming traffic for fear of having his image captured on a dashcam.[23] People are more easily recognized face-on than in profile. That short observation from his book *Permanent Record* delineates some salient themes in the cryptographic city: surveillance, risk and paranoia.[24]

What he and other insiders exposed was that the NSA was able to obtain wholesale communications data from every citizen on the phone network in the United States and abroad. The initial revelation provided by Snowden to *The Guardian* and published on June 6, 2013, stated that for telephone communications the NSA covert operations accessed the phone numbers of both parties at either end of a call, with location data, unique identifiers, and the time and duration of calls.[25] The NSA was not requesting from the communications service providers the content of conversations, but the *metadata*. We normally think of secret service investigations as directed at key targets, but here the data of everyone on the telephone network was collected whether or not they were suspects. The data was stored in bulk on vast servers ready to be mined as needed. Software could trace links and detect patterns. The collection of metadata was automated, and no human being needed to ever see the data—unless authorized to investigate particular individuals. Here the security agencies obscured their engagement with the data by the claim that it is not the data they want to inspect but data about the data.

Snowden and others have argued that a great deal about a person's life can be harvested from such metadata, including networks of contacts, lifestyle, activities, and competencies. This information space is even more revealing if you include the bulk collection of email metadata, browser histories, debit card transactions, and travel data, especially if linked. Is it really surveillance if the metadata is collected in bulk and not inspected by a human operative?

A helpful blog post by information law lecturer Paul Bernal explains the problem of identifying when surveillance actually happens. There are three key moments: "the gathering or collecting of data, the automated analysis of the data (including algorithmic filtering), and then the 'human' examination of the results of that analysis of filtering."[26] Does surveillance happen when the bulk data is collected, or when humans inspect the data?[27]

Bernal argues that the same privacy question arises in the case of video surveillance: the moment the surveillance system is installed, when there's the
means for someone (a relative, the landlord, an employer, a law enforcement official) to see what the camera sees, even if they never take up the
opportunity.

Urban Vulnerabilities

The risks caused by cyber espionage are consequential for cities and communities. The NSA headquarters is located between the cities of Washington, DC, and Baltimore. Among its many operations the NSA develops
digital tools for spying on other countries and exploiting some of the
vulnerabilities I have described. *Wired* magazine reported that "US Cyber
Command has penetrated more deeply than ever before into Russian electric utilities, planting malware potentially capable of disrupting the grid,
perhaps as a retaliatory measure meant to deter further cyberattacks by the
country's hackers."[28] But some of these tools leaked out, and around 2017
were turned on the city of Baltimore to cripple its infrastructure. According
to a *New York Times* report, "For nearly three weeks, Baltimore has strug-
gled with a cyberattack by digital extortionists that has frozen thousands
of computers, shut down email and disrupted real estate sales, water bills,
health alerts and many other services."[29] Many NSA workers live in Baltimore, so the NSA's hacking and counter-hacking operations rebounded
onto those citizens.

The metaphor of *urban vulnerability as commodity* offers an interesting
lens through which to consider city challenges. The nefarious exploitation
of information about vulnerabilities that might result in failure of some
kind recalls persistent urban scenarios of protection rackets, black markets,
and insurance scams. Failure by a sports player is worth something to bad
actors who rig games and sports that involve betting. Traders in stock markets benefit from insider knowledge about impending failure. In the worst
cases such failure can be engineered, with or without the complicity of the
protagonists.

As another disreputable practice, political parties and interests can put
forward "spoiler" candidates who they know will fail to be elected, whose
policies accord with some opposition voters and effectively splits them
away from their support for the mainstream opposition. These candidates

are encouraged to run and are installed to fail and to dilute the opposition's vote.

There are also brokers who trade portfolios of failed urban enterprises, failed retail outlets, and unprofitable property investments. These put losses on the ledger and provide a way of avoiding corporate taxes in some countries. Disreputable competitive practices may induce failure in competing enterprises to facilitate takeovers. Once written off, products that fail in terms of profitability can be used to bulk up product portfolios and serve as lures within bargain offerings. Under this framing, wealth disparity, homelessness, and uneven access to technical and social infrastructures also constitute urban vulnerabilities readily exploited by negative and disruptive political agents—not to mention opportunistic agents who seek to profit from pandemics and other calamities.

I began this chapter with the tactics of obfuscation that are means for concealing data and fall within the repertoire of urban tactics deployed by state instruments, corporations, activists, and day-to-day providers and consumers of information. That led me to consider the wider challenges presented by cyber espionage, cyberattacks, ransomware, and other threats to urban citizens and infrastructures.

Personal data circulates in the cryptographic city. Algorithms can calculate inferences about individuals, groups, and whole populations from such data. For example, pattern-matching algorithms can categorize people according to the words they use in their social media posts.

Researchers at the University of Cambridge Psychometrics Centre developed a demonstrator program that claims to derive your personality profile, age, and other information from how you use language in a blog, Twitter feed, or email.[1] Anyone can interact with the demonstration program on the Apply Magic Sauce: Trait Prediction Engine website. Calling on machine learning techniques the researchers see this demonstrator as "a modest attempt to reverse the trend in Big Data and empower citizens to not only retain control of their data but also derive meaningful insight from it." The program "learns" from an extensive training set of texts and author profiles. It then estimates the personal attributes and psychological profiles of the authors of new texts. It calculates whether you are likely to be conservative or liberal, impulsive or organized, competitive or trusting, relaxed or succumb easily to stress.

The Cambridge program demonstrates the potential of text and other data harvesting. Pattern-matching algorithms can take the content of our social media posts and even our sequencing of keyboard strokes to make inferences that put us into categories as individuals or groups. The idea is that our digital footprint reveals more about us than we state explicitly, and some of it can be gleaned not only from what we write, but also from actions such as our choices of images we put on Instagram, TikTok, or Snapchat. From such incidental data, machine learning might estimate our educational attainment, ethnic background, social circle, disposable income, purchasing habits, the kinds of holidays we take, and alcohol consumption.

Your digital footprint also provides a rough guide to your politics. Such systems are not always accurate in assessing personality or background, but they don't have to be. They serve to narrow the target for public information, advertising, propaganda, and social interference campaigns to exert influence or disrupt the opinions and habits of citizens. Accurate, individual personal profiling from your digital footprint is error-prone, but it is the aggregation of such assessments across whole populations that provides the benefits to advertisers, persuaders, and political actors as outlined in chapter 11. If political campaigners can target enough people with a message tuned to voters' social and personality profiles, then that could be sufficient to tip a vote in favor of one candidate over another.

With this kind of personality profiling, social media platform developers can also fine-tune their systems to maximize revenue by generating controversy and hence engagement. The release of the trove of papers ("The Facebook Papers") by whistle-blower and former Facebook employee Frances Haugen in 2021 suggests that Facebook presents its newsfeed content to polarize opinion and keep people engaged, if not "addicted," in their social media news feeds. At a US Congressional hearing, Haugen asserted, "I'm here today because I believe Facebook's products harm children, stoke division and weaken our democracy."[2] The charge of harming children related to making "young girls and women feel bad about their bodies."[3] The reference to democracy suggests the programs that configure your newsfeed on Facebook tilt what you read toward controversial posts authored by readers and contributors, further influencing people's politics. The "active measures" outlined in chapter 11 can also exploit such methods. That's part of the opportunity, challenge, and peril of big data in the cryptographic city.

Where they occur, these monitoring processes are covert, involving company policies that are hidden from users, but also involve algorithms that are little understood by people who use the platforms. Similar hidden processes are at work in online retail and booking systems. We become aware that some calculation is happening in the background of our interactions when we receive directed advertising, or the platforms appear to come preloaded with preferences derived from our purchasing history on this or other platforms.

Online retail is a key instance of the cryptographic economy and relies on methods for hiding and securing transaction information. Such processes also rely on their own hidden protocols as they interface with users.

Shoshana Zuboff exposes how these patterns provide information about consumer behavior, much of which constitutes a hidden commodity traded and sold between companies, often without our knowledge or explicit consent.[4]

Algorithms Everywhere

Processes of online selection, booking, and purchasing are at the interface of the platform experience for many urban consumers and Internet users. So far, I have invoked terms such as *programs, systems, applications, apps, platforms,* and *algorithms* as the vehicles for these hidden data-harvesting processes. The word *algorithm* pertains to information processing via computer, as "a precisely defined set of mathematical or logical operations for the performance of a particular task" (OED). The word shares its derivation with *algebra,* which is a term I am prepared to associate with a specialized knowledge and was a significant and intimidating aspect of my own schooling. Social media commentators have directed their attention to the "algorithm" as a target of concern. For example, in her recent study into digital technology Ruha Benjamin poses the challenge, "What do 'free will' and 'autonomy' mean in a world in which algorithms are tracking, predicting, and persuading us at every turn?"[5]

It is worth examining what algorithms mean to the cryptographic city. To elaborate on the OED definition, algorithms are precise and repeatable procedures for accomplishing a computable task. An algorithm is a component of a computer program directed at a particular subtask and with an identifiable logic. So there are algorithms well known to programmers for searching a text file or database for the occurrence of a particular word, and algorithms for compiling and sorting lists of words, and processing decision trees.[6]

It is a commonplace to remark that algorithms are typically "hidden." They are "black boxes," the content of which may be known only to the author of the algorithm. In fact, computer programmers typically draw on libraries of algorithms to assemble their own algorithms. Some of these library elements are specific to the operating system or the brand of microchip the computer program is running on. In so far as *algorithm* serves as a useful term to describe what happens in digital systems, algorithms are combined, configured, and nested and are transparent and visible to varying degrees in their development and deployment and for different programmers, engineers, developers, and users.

Algorithms are responsible for encrypting and decrypting files and data flows, and create, process, and compare hash strings. Algorithms also activate the pixels on a display screen to form text and images, open and close files in response to mouse clicks, manage and transmit bit strings through networks, and perform countless benign operations that any computer user is unlikely to know or care about.

The algorithms that draw opprobrium from some critics are likely those that surreptitiously monitor our purchasing habits, mouse clicks, screen attention, and flows of social media data. Whether they implement covert social media monitoring or enable web search, algorithms and their combinations inevitably embody values. I have referred to Ruha Benjamin's commentary on algorithms. Her book is really about racial biases in the digital realm: *Race After Technology*. She writes, "It is certainly the case that algorithmic discrimination is only one facet of a much wider phenomenon, in which what it means to be human is called into question."[7] Appropriate and inappropriate bias is impossible to eradicate, though fair and open social discourse and action requires that we are always prepared to challenge our biases.[8]

Bias is common in any technology. As an urban example, until special interest groups lobbied regulators and designers to make buildings more accessible, buildings would limit access to people with a particular range of physical mobilities. Architects realized that the width, swing direction, threshold, and opener of the ubiquitous and ordinary office doorway could include or exclude people who would otherwise need to use the door.[9] Prior to that awakening, and legislation, a value system was in play behind the affordances of doorway design that many designers would take for granted. So too, the design of a smartphone assumes a certain dexterity and visual acuity that able-bodied designers take for granted. Interaction design also assumes certain value systems grounded in assumptions about cognition, consumption, communication, and sociability.

Bias is evident in the algorithms, but also the data, the structuring of the data, and what the algorithm admits or excludes as data in the way it is designed. In some cases bias is coded into weightings attached to different data components. As a further urban example, the UK's Consumer Data Research Centre provides data and maps to indicate where residents are poorly served by public facilities and where incomes are low, in other words, areas of multiple deprivation.[10] The algorithms process and display relevant data about income, numbers of people in households, and distances from amenities.[11] Controversy arises about the respective weighting applied to each of these factors in identifying and mapping areas of poverty. The weightings inevitably and explicitly encode biases, and different weightings produce differing distributions of deprivation on the maps.

Who is responsible, and therefore accountable, for what an algorithm does? Who feeds the values into the design of an algorithm? In the case of a building design there are many "authors," secreted within the value systems of manufacturers, suppliers, consultants, owners, regulators, funders, professional bodies, and educators. So too there are many authors of a typical computer program. In her book *Cloud Ethics*, Louise Amoore asserts that "the algorithm already presents itself as an ethicopolitical arrangement of values, assumptions, and propositions about the world."[12] This multiauthor view informs her account of "algorithmic ethics."[13] We want to identify the human agent responsible for biases, errors, and inequities we encounter online. Yet responsibility resides with a multitude of agencies.[14] By this reading, the focus on algorithms as the purveyors of hidden bias is an attempt to identify people and things that are ethically accountable.

Programmers write algorithms to disrupt or confound the functions of other algorithms, and to break through cryptographic defenses as in the case of the espionage measures described in chapter 11. In that chapter I also described algorithms and methods that deploy obfuscation as a means of confounding the operations of platforms, infrastructures, and even social organization. These tactics can operate in service of either the predator or the prey, the aggressor or the victim.

On the side of the "prey," agents that seek to protect consumers have devised obfuscation tactics to confound data harvesting. The app called AdNauseam disrupts the operations of web platforms that profile you according to the ads you click. When installed in your browser the app automatically sends clicks to the server for every ad on a page. According to the adnauseam.io website: "As the collected data gathered shows an omnivorous click-stream, user tracking, targeting and surveillance become futile."[15]

Advertisers and profilers also want to know where you are. To confound this locational information the TrackMeNot browser plugin sends false navigation data to the server that is trying to surveil you. It obscures your locational coordinates. TrackMeNot blends fake and actual search data to obfuscate profiling. The trackmenot.io website says, "With TrackMeNot, actual web searches, lost in a cloud of false leads, are essentially hidden in plain view."[16] In their book *Obfuscation: A User's Guide for Privacy and Protest* (referenced in chapter 11), Brunton and Nissenbaum offer a range of consumer-oriented obfuscation tactics.[17]

Feature Variables

The average user of a computer program will not know or care about the myriad algorithms it deploys. The whole program is hidden as are its functions and procedures. Algorithms make use of slots in computer memory into which are stored numbers and characters as binary coded variables. All we consumers see is the interface and a subset of variables relevant to our interactions.

Some variables exist on a numerical scale, identified as quantities: age, SAT score, income, house number, property area. If I only know the dimensions of my property, then the area is a variable that is hidden until I calculate it from the dimensions. However, some variables are more hidden than that.

The identification and influence of hidden variables is one of the major challenges of statistical and machine learning methods deployed in big data analysis. Unknown factors may lie latent in the data as "confounding variables." That is another category identifying variables that are invisible because we do not have precisely the right information that brings them to light. Techniques for clustering concepts serve to identify and instantiate the values of variables. These include statistical analysis and neural network methods of machine learning.[18]

Considering the high premium placed on the visual affordances of the built environment, automated feature detection offers high rewards for organizations. Amazon offers a service for businesses to identify features in large collections of images. According to the aws.amazon.com/rekognition website, the service provides automatic labeling of elements in a picture (e.g., here is a person on a bike, there's a traffic jam). A company can identify if its brand label happens to appear in a news report, identify image rights violations, detect inappropriate or dangerous content or objects, recognize celebrities, and check whether people are wearing the right personal protective equipment (PPE).

The Google Cloud Vision AI platform provides an API (application programming interface) for researchers to identify features within large numbers of publicly visible images, as on photo-sharing platforms (Flikr, Snapchat, Instagram) or an individual's own private image collection.[19] That has potential uses in detecting what people focus on as important, attractive, interesting, or "instagrammable" about a place. Several platforms offer similar capability. The features detected are typically words (house, tree, sky, etc.) with a confidence number attached. Unlike the tags or metadata people sometimes attach to their digital photos, the platform's algorithms generate these as feature lists automatically.

Microsoft Word provides automatic "alt text" creation based on features for image content that text-to-voice readers can recite for people with visual impairment. Figure 12.1 shows an urban image and the features Google Cloud Vision API plugin detects automatically, along with a confidence ranking. With this plugin, users of the software can select the features they want to adopt as tags to assist in search at some later date, or as the basis of their own alt text descriptors.

Google generates feature tags based on its access to very large stores of online imagery in which photographers, users, or human operators have already identified the content.[20] These word tags help the search algorithms identify and filter images as well as match images that have similar features. The current incarnation of the Google Image search facility (Google Lens) on a smartphone matches locations as well as images and returns links to relevant websites. The app will even identify species and types of animals, plants, and human-made objects. The platform also delivers collections of similar images that have also been tagged automatically with such features. Automated feature detection is accessible to anyone with a networked computer or a smartphone.[21]

Feature detection within images deploys various machine learning techniques. As I have already suggested, a machine learning algorithm scans thousands of "training" images that are prelabeled with relevant feature descriptors. The algorithm adjusts the numerical variables in its network data structure to reproduce those same labels when it encounters those images again. It thereby "learns" to re-identify those features. More important, the variable adjustments are such that the algorithm can detect the

Labels

Label	%	Label	%
Cloud	97%	Monochrome Photography	76%
Sky	96%	Façade	75%
Building	95%	Crowd	75%
Daytime	95%	Human Settlement	74%
Window	93%	Pedestrian	74%
White	92%	Mixed-use	73%
Black	90%	Street	71%
Infrastructure	89%	Downtown	70%
Street Light	87%	Travel	69%
Style	84%	Winter	68%
Urban Design	82%	Walking	68%
Tree	82%	Event	68%
Neighbourhood	82%	Street Fashion	67%
Public Space	82%	Spring	65%
Road Surface	82%	Stock Photography	63%
Road	80%	Apartment	62%
Sidewalk	80%	Tourism	61%
City	79%	Town Square	61%
Thoroughfare	79%	Cobblestone	58%
People	78%	Recreation	57%
House	78%	Commercial Building	57%
Metropolitan Area	77%	History	55%
Monochrome	77%	Transport	54%
Metropolis	77%	Plaza	52%

Objects

Object	%
Building	90%
Person	89%
Building	88%
Building	87%
Building	79%
Clothing	77%
Person	77%
Person	76%
Person	74%
Person	73%
Person	73%
Person	71%
Person	64%
Clothing	61%
Person	59%
Person	51%

Figure 12.1

Feature detection in a photograph. Tabulated lists are from the Google Cloud Vision API which also identifies and outlines areas in the image to which the tags apply. The Google Lens app delivers similar information, including a collection of pictures from the web that are similar. Automated text generation by Microsoft Word provides alt text: "A group of people walking on a sidewalk next to a street. Description automatically generated with medium confidence." Photograph by the author.

same features in new images it has not previously scanned. This is a neural network approach to machine learning. Amazon describes its feature (object) detection algorithm as one such "deep neural network": "It is a supervised learning algorithm that takes images as input and identifies all instances of objects within the image scene. The object is categorized into one of the classes in a specified collection with a confidence score that it belongs to the class."

The "learning" process is not entirely automated. Neural network developers have to devise the network configuration, decide what constitute inputs and outputs to the network, the layers in between (hidden layers), and the sensitivity of the network's numerical variables—weightings, probabilities, network connections (edges), and threshold values.

Louise Amoore explores the ethical dimensions of designing a neural network: "This spatial arrangement of probabilistic propositions is one of the places where I locate the ethicopolitics that is always already present within the algorithm. The selection of training data; the detection of edges; the decisions on hidden layers; the assigning of probability weightings; and the setting of threshold values: these are the multiple moments when humans and algorithms generate a regime of recognition."[22] She is critical of the reductive nature of feature detection. In any case, though they are effective in processing large numbers of images, automated feature detection algorithms are less accurate or nuanced than human beings. The algorithm misses the "features" in figure 12.1—that there's a marquee in the frame, a glass-roofed atrium, or that it is windy—and is incapable of delivering the meaning and significance of the picture as a whole to its various human interpreters and audiences.

Automated feature detection in imagery is a useful test case for the ethics of machine learning, though the application of the techniques extends to other sensory modalities as evident in surgical procedures that include multiple sensory skills such as touch and precise movement. Amoore makes the case that machine learning highlights the conflicted nature of attributing responsibility. Who is responsible for errors and misjudgments? Mistakes in robotic surgery procedures, errors in automated drone strikes, harm to noncombatants and false matches in image analysis provide obvious examples. She suggests that the surgeons, drone operators, and photographers who create the training sets on which the machine learning is based carry some responsibility. I would argue that the ethical responsibility extends to those who design, select, adjust, and deploy the learning algorithms.

The argument about shared and conflicted attribution is similar to discussions about authorship, originality, and intellectual copyright within the creative professions. I agree with Amoore that attribution is contextual, and fraught, and is resolved by human judgment: "Ethicopolitical life is about irresolvable struggles, intransigence, duress, and opacity, and it must continue to be so if a future possibility for politics is not to be eclipsed by the output signals of algorithms."[23]

Probabilities

Machine learning algorithms that match images of the kind I have just described typically provide sets of features detected with a confidence or probability value attached as in figure 12.1.

Statistical information also helps establish the probability that a particular event will occur, such as the probability that any individual will be involved in a traffic accident. For example, the probability that a traffic accident is due to driver carelessness is over 50 percent according to some studies.[24] Statistical analysis can also establish that one event followed another, so that the probability that a traffic accident will be followed by hospitalization of the accident victims might be 12 percent. Algorithms that control traffic lanes or self-driving cars can take such events and sequence probabilities into account to make predictions. The class of computational techniques for this kind of calculation are the Hidden Markov Models (HMM) referenced in chapter 4, with applications ranging from managing urban mobility to gene prediction.[25] Some automated speech recognition and language translation programs also deploy HMMs to process information about word sequences. To the technologies of writing and print in chapter 2, we can add speech-to-text, text-to-speech, and automated language translation as factors in the operations of the cryptographic city.[26]

Andrey Andreyevich Markov (1856–1922) was a mathematician who devised methods for modeling probabilistic processes. Sequences of events linked by probabilities in the way I have described is called a Markov chain. That is a chain of events for which the probabilities of any event A (such as a traffic accident) will be followed by another event B (hospitalization) irrespective of what events preceded A.

Generalizing this method, the nodes in a network diagram could be decision points, such as navigating from one web page to another, or road junctions encountered while a driverless car navigates through the city. The links in the Markov chain network are labeled with the probability that the driverless car would take that path given statistical data about congestion or road gradient.

Markov modeling is a further example of information "hidden" within data that exploits probabilities and informs operations in the cryptographic city. The second element in the algorithmic armory of big data is to exploit the repetitive nature of much human and machine actions, and of data flows that repeat.

Coefficients

There exists a class of urban problem-solving that reduces complicated variations in data values to a series of simple cyclical patterns of different frequencies. The method determines the contribution of each of those frequencies to the overall pattern in the data. The process identifies the main frequencies and a series of coefficients that show the percentage of each frequency in complex data as explored in chapter 10. That is a useful type of analysis as it helps tease out patterns in data and thereby identify variables hidden within it. One such method is known as discrete cosine transform (DCT).

A key article by Ahmed et al. on the DCT method presents it as a means of recognizing patterns in data.[27] In figure 12.2 I show a simple experiment in which I attempted to capture regular cycles in publicly available local data about COVID cases at the early stages of the pandemic. The automated process attempts to match a series of repeating cosine curves to the data. The result is a series of coefficients, numbers that represent how much each cosine curve contributes to the overall wave pattern in the data. The DCT method also serves as a way of compressing complex data, as you only need to store the coefficients and simple indices that designate each cosine curve. The data here is in just one dimension, showing numbers of infections over a short time period. This is a demonstration and founded on the assumption that there are time-based regularities that contribute to the COVID case count, such as people's weekly activity cycles, alterations in the virus, waves of resistance, immunity, other diseases, other

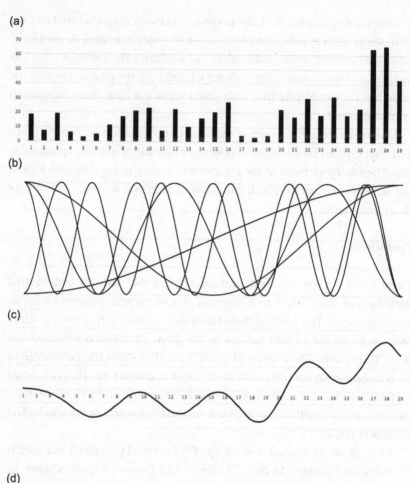

Figure 12.2
Discrete cosine transform (DCT) analysis of the number of people who reported COVID-19 each day in Scotland over a twenty-nine-day period in March 2020. (a) Numbers of cases over twenty-nine days. (b) The five most prominent component frequencies detected in the data by DCT adjusted to the same amplitude and with smaller frequencies filtered out. (c) Reduction of the original data to its major frequencies. (d) The most prominent frequencies and their contributions to the data as coefficients. *Source:* Author.

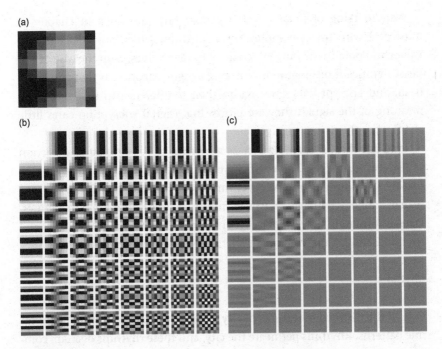

Figure 12.3
Reducing the color frequencies in an 8×8 target image (a). The second image (b) shows 64 frequency variations in gray across an 8×8 array of pixels. The variations follow simple cosine distributions at different frequencies in the x and y directions. The 2D discrete cosine transformation (DCT) method effectively tests each of the 64 8×8 variations in the array (b) against the target image (a) to detect the percentage of their presence. The second 64 variations array (c) shows the presence of each frequency as variations in luminosity. When overlayed, each of the 8×8 pictures in the second array would approximate the target image. The frequency variations that appear mostly in gray contribute little and can be filtered out when the file is saved. JPEG compression uses this method to reduce file size. Cryptographers can use the gray areas to store hidden information as a steganographic technique. *Source:* Author.

environmental responses of the virus, or the introduction of new vectors for transmission.

The DCT method can also be applied to 2D information, as in the case of mapped data or pictures. DCT helps to analyze or compress picture images (figure 12.3). As outlined in chapter 10, the JPEG method of image compression uses DCT coefficients to filter out inconsequential variations in color intensity across an image. It breaks an image into 8×8 pixel squares and stores the major coefficients for each square.

Watermarking of images and steganography (described in chapter 9) must deal with image-compression algorithms, which adjust pixel color values and potentially purge the data of hidden values, as in the case of LSB (least significant bit) methods of hiding images. Algorithms that compress, hash, and encrypt data converge in their indifference to the content or meaning of the signals they are processing. Digital encryption cares little about what the data means.

I include this 2D example here as it pertains to the automated detection of features in images. The coefficients are hidden features, scarcely recognizable by a human observer, yet important for algorithms that process, search, and match images, as well as those that hide information in pictures. Data, information flows, and images harbor hidden frequencies, cycles, and rhythms. Similar analysis reveals rhythms hidden within the city.

Rhythms

One of the distinctive features of city living is the concentration of rhythmic patterns. Rhythms permeate the city, and these rhythms overlap, combine, aggregate, and interfere with one another. That is the gist of Henri Lefebvre's (1901–1991) book entitled *Rhythmanalysis*.[28]

By my reading the concept fits within the genre of research concerned with *everydayness*, the *quotidian*, which implies a concern with ordinary things and everyday phenomena that repeat. Lefebvre's book first appeared in 1992 before the boom in big data. Few commentators at the time saw any potential for computers to do justice to the ordinary and everyday experience of repetitions, cycles, and rhythms.

As I explored in my book *The Tuning of Place*, mundane and ordinary events are also everyday events, that is, events that occur every day, repeatedly, and relate to people's habitual activities.[29] So, a rhythmanalysis will focus on ordinary events and things that we take for granted. The book *City Rhythm* by Caroline Nevejan, Pinar Sefkatli, and Scott Cunningham recruits rhythm to explain cities and their internal diversity, as well as to account for differences between cities.[30] Their book also introduces the application of HMM as a way of explaining what happens in cities. *City Rhythm* amplifies the rhythm metaphor by mapping what the authors refer to as "beats," "base rhythms," and "street rhythms" across cities in The Netherlands.

I think by "beats" the authors mean the dominant rhythms of a region, exemplified by the ebb and flow of the volume of pedestrian and road traffic. By their reading, base and street rhythms "show significant transitions over time for the specific area."[31] I'm interested in the authors' claim that this kind of analysis can bring out similarities and differences between regions. According to the book, rhythms influence how people feel in each other's company: "When sharing rhythm, people feel more at ease with each other."[32] Rhythms engender trust: "When recognizing each other, people synchronize and tune their rhythm to each other."[33]

The rhythm concept also helps explain the mismatch between citizens and the systems of the city, in particular the road system: "The roads are too wide and busy, and the traffic lights are too short to cross the streets. This situation is reflected on the mismatch in the rhythms between the elderly and the rather fast rhythms that the neighbourhood presents."[34] Such rhythmanalysis helps explain conflicts between city inhabitants, as in the case of the insecurities felt by some less-mobile people when in proximity to exuberant youths. One image in the book shows a series of frequency curves, approximating cosine curves, of the intensities of activities across a typical week in the Keizerswaard shopping center.[35] Drawing on Giles Deleuze and echoing some of the ideas I explored in *The Tuning of Place*, Nevejan, Sefkatli, and Cunningham show how: "A territory happens when different rhythms come together and they create their own expressive language."[36] Rhythmanalysis is an example of semi-formal algorithmic analysis applicable to the cryptographic city. The hidden variability of city living is temporal as well as spatial.

Secret Synchronies

I will conclude this chapter by reinforcing the relationship between rhythms and patterns of secrecy in the city. Patterns in cycles are among the panoply of hidden dimensions in the city. The methods I've been describing illustrate a basic truism: events follow one another in sequence, inexorably and in daily, weekly, monthly, and annual cycles. At a personal level, you brush your teeth, you wash your face, you pour the cereal and milk, you eat it, you rinse the bowl, you attend an online meeting, you get dressed, and so the day proceeds. Such sequences follow patterns. In some cases, an automated system might attempt to detect those patterns: to predict what comes next, to show how to reinforce or break out of a pattern, or to detect the variables that influence those event sequences. We often describe deviations from the norm in terms of cycles. It is a time-worn conception about creativity: to think out of the box you need time out of the schedule, the routine, the humdrum world of everyday matters.

The concept of urban cryptography as a commerce in hidden messages has never been far from my considerations in this chapter. Some writers define the act of keeping secrets with recourse to cycles and frequencies. For example, an online *Psychology Today* self-help article explains that to keep a secret is a "habit of mind," and it is mostly a bad habit—something to break out of.[37]

But to investigate secrets does not require us to make such a judgment. An academic article on "family secrets" by Mark Karpel refers to the "loyalty dynamics in the creation, maintenance, and eventual facing of secrets in families."[38] Secrets require maintenance—repeatedly. Not only are some secret acts performed in repetitive cycles, but secrets of any consequence encounter repeated onslaughts from exposure, require repeated evasion and denial, and yet more elaborate and repeated cover-ups and obfuscations.

We've seen in public forums that secrets are often in the company of lies, and lies are rarely singular, but get repeated, amplified, and woven wickedly into sticky webs. Secrets and lies are habituated exercises in repetition. I first made these observations during the tenure of the forty-fifth U.S. president when commentators would note the repetition of the president's mistruths: the size of the inauguration crowd, that the conversation with the Ukrainian Prime Minister was a "perfect call," that the election the president lost was fraudulent. Lies and propaganda have to be repeated to reinforce them and to resist the weight of contradictory evidence. Such repetitions also make use of weekly and seasonal news cycles. Active measures, infiltration of social media news feeds, and kompromat by foreign adversaries exploit such cycles and repetitions.

I take it for granted that events repeat or fit into a cycle of repetitions. Whether or not they are made explicit as schedules, our everyday lives are permeated with events that occur every day, or every other day, or weekly, annually, hourly, by the lunar calendar, the seasons, tides, breaths or other cycles, and cycles within cycles. Agents of such occurrences may wish to expose events to different constituencies, or hide them altogether, as is the case when the social media user adjusts the privacy settings for particular postings.

How do you maintain secrets so that they are invisible to others on a cyclical timeline? One method is to resist other people's cycles, simply by operating acyclically or asynchronously. If you've shared a home with someone who goes out in the evening and is back in bed just as you are getting up, then you'll know the relationship between cycles and secrecy.

As I've seen in heist and prison escape dramas, those working in secret operate counter to regular patrols, the sweep of the surveillance camera or spotlight, opening hours, night porter duties, and the usual daily cycle. Habitual criminality weaponizes the counter-schedule. Seasoned criminals work on different cycles from their victims, and the dark arts of cyber criminality adapt to the complex periodicities of 24/7 global communications.

One of the challenges in detecting crime, and the same applies to uncovering secrets, is to know if the events you observe are part of a concerted plan or merely coincidences. I think of a coincidence as the meeting of two or more cycles at a particular moment. Secret deeds give themselves away through coincidence. That is one of the ways to catch someone in a criminal act—align the patrol with the cycles of the criminal.

In his *A Burglar's Guide to the City*, Manaugh refers to homeowners' "rhythms of vulnerability."[39] Most homeowners know about the risks of leaving a place unoccupied and have simple electronic timers that turn the lights on in the evenings while they are on holiday. You can even install a flickering light source that suggests someone has the television set on.[40] Then there's the practice of "oversharing" on social media, by which burglars can easily deduce whether or not you are at home, and sometimes infer the location of your home. There is (or was) a website to test such vulnerabilities called pleaserobme.com.

Cyclical properties of algorithms, variables, probabilities, and coefficients constitute hidden dimensions to the cryptographic city. These hidden features also make worlds, universes, and cities beyond our ordinary apprehension, the subject of chapter 13.

13   Hiding in the Multiverse

In 2021, Facebook and other technology companies committed to developing the "metaverse," "a set of virtual spaces where you can create and explore with other people who aren't in the same physical space as you."[1] People with mobile phones and who communicate via video link are accustomed to "being with" someone who isn't with them in person.[2] But to frame this experience in terms of the metaverse implies that it is possible to occupy a realm other than the world of everyday embodied experience. The evasiveness of these metaverse experiences entitles us to think of them as yet another aspect of the cryptographic city. As I will show, cryptography enters the discourse as writers and scientists speculate about communications between such covert places. Theoretical quantum physics adds some credence to the existence of multiple worlds. That leads me to quantum computing, the prospect of which impacts the way we think about encryption.

Hidden Dimensions

A book from the 1960s by Edward Hall bears the promising title *Hidden Dimensions*. The term *hidden dimension* is alluring for the cryptographic city as it implies some properties of spaces beyond ordinary perception. The book is about *proxemics*, the study of how people organize and use space in their interpersonal communication. Hall indicates what he means by hidden dimensions as he explains the spatial arrangement in an office. It consists of three zones: "1. The immediate work area of the desktop and chair. 2. A series of points within arm's reach outside the area mentioned above. 3. Spaces marked as the limit reached when one pushes away from the desk to achieve a little distance from the work without actually getting up."[3]

He advocates that designers take account of the fluid and practical dimensioning of space that goes beyond mere room dimensions. These are the dimensions of movement defining "kinesthetic space,"[4] an important element of people's everyday experience. Hall draws attention to certain social practices that pertain to such hidden dimensions. These are less obvious than the visible and *observable* sizes and shapes of rooms and need to be inferred or otherwise discovered. A dimension is simply a measure, and here a *hidden dimension* is a measurement of the space formed by someone's movements, distances they stand from other people, and spaces defined by where occupants place objects, for example, things you put close (your laptop, personal photos, reminder notes, your lunch) and objects that can be further away (reference books, the printer, old files, armchair).

These spatial affordances are also contingent elements of a so-called subarchitecture[5] and are apprehended in the context of practicalities and various sensory modalities and can be designed for, interpreted openly, and finessed. Theorist of sound and space Brandon LaBelle writes that subarchitectures admit into consideration "the minor practices of space."[6]

The concept of *dimension* normally equates to a gridded world of three dimensions along x, y, and z axes, thanks to Descartes' definition of objects in space by the variables of length, breadth, and height. A one-dimensional world is defined along a line, just one axis of the grid. Two dimensions involve a world as drawn on a flat sheet of paper (xy).[7] The introduction of a fourth dimension is less accessible to direct human experience. Some think of the fourth dimension as *time*, though you cannot move backward and forward along a time axis as you can the three primary dimensions.[8]

Descartes was a mathematician and pioneered methods in geometrical algebra and calculus. We can also thank Descartes for identifying and naming an "imaginary" number, i, which is the square root of -1. And thus developed methods for plotting impossible numbers on the complex number plane. Those multicolored filigree images you see in online videos if you search for "Mandelbrot zoom" have at their core a simple looped algorithm to visualize the contoured fringes of this infinitely varied self-similar fractal space.[9]

Dimensionality is a mathematical concept, and there are many attempts to visualize n-dimensional and complex number space on a flat computer screen, and with stereopsis in virtual reality. The basics of geometry, visualization, and extrapolation lead people with imagination to speculate on "hidden dimensions" that evade perception by the usual exercise of the human senses. To pick up on another geometrical metaphor some scholars and fantasists think of *parallel* worlds.

These imaginative spatialities seem to defy rationality. As they cannot be perceived via the usual sensory channels, they epitomize for some people the idea of a hidden place. From a psychoanalytic perspective we could say these supposed hidden dimensions are mental surrogates, something we can discuss in a detached manner, without probing into the inner recesses of human minds where real secrets reside. It is easy to dismiss such places as those "of which we cannot speak,"[10] but imaginative concepts leak though into the practical everyday domain. They inform, test, and challenge concepts in our lived realities. They imply entry to and exit from a place that requires a secret code, a key, a passcode. For the time being I'm content to concede, or at least speculate, that such imagined realities are embedded into the fabric of the urban lifeworld.

A Multitude of Multiverses

Gardens and cities provide the two main models of places beyond everyday experience. There are the Garden of Eden and the Celestial City. Italo Calvino's *Invisible Cities* presents each of any city's characteristics and infrastructures as if a separate city, of water supply, internment, drains, or signage.[11] Darran Anderson's *Imaginary Cities* collates further imaginative features of cities.[12]

Many terms come to our aid in identifying such places. A search online for the term *multiverse* returns several variants: parallel universes, alternate universes, interpenetrating dimensions, parallel dimensions, parallel worlds, multiple worlds, parallel realities, alternate realities, alternate timelines, alternate dimensions, dimensional planes, quantum universes, and quantum realities.[13] I would add some further variants: planes of existence and hidden dimensions.[14] Numerous science fiction, fantasy and speculative novels and films play on the multiverse theme. The *Flatland* story by Edwin Abbott published in 1884 tells of commerce between worlds each defined by one-, two- and three-dimensional geometry.[15] Philip Pullman's book *Northern Lights: His Dark Materials* taps into a multiverse of realities.[16] The book opens with a quote from John Milton's *Paradise Lost*, in which the poet speculates that the creator might "ordain His dark materials to create more worlds." There are multiple multiverses. I have already alluded to the "hidden dimensions" of our practical, everyday lifeworld. Here is my attempt to tease out some further distinctions. Multiple universes appear in narratives about framing, altered states, ancient concepts dividing the material and the ideal, multiverses in theoretical physics, and geometry. I'll review these briefly.

Alternative universes feature in the identification of delusion and dishonesty. Commentators frequently described the forty-fifth U.S. president as living in an alternative universe, variously described as immaterial, parallel, imaginary—conjuring up a universe of misinformation.[17] This style of critique points to an alternative framing of reality, to a plurality of points of view, different interpretations, framings, lenses, through which we each see the world according to our own perceptions, preconceptions, biases, and expectations. An authoritarian will say only their particular framing is valid. Contrary to that fixed position lies a naïvely relativistic "multiversality" in which we think anyone's view of reality is as valid as any other. Either extreme will be dominated by the affordances of the world views posited by the loud and powerful.

A more prosaic expression of multiversality focuses on altered states. To transition from one dimension to another is to change from one state to another. Water changes from vapor to liquid by lowering the temperature. If this transition is about a shared universe of human experience, then altered states are as common as changes in the weather. If we are referring to psychological or emotional states of the human being then there are therapies, stimulants, and depressants to assist, or a change of scene, exercise, company, or a good night's sleep.

The legacy from which multiversality draws relies on long-standing distinctions between the physical and the spiritual, the material and the ideal, earth and heaven, waking and dreaming, conscious and unconscious, real and imaginary—each expresses long-standing alternative realities. Here, "alternative" is the correct term to use, as these are alternating, binary distinctions and "universes." I am enough of a structuralist to recognize the complex priorities involved in the formulation of such binaries. The Neoplatonists articulated hierarchical orders of reality. Some states, worlds, or conditions are more real, virtuous, wise, or enlightened than others. I would also group fantasy and science fiction parallel universes within this category. At least, these genres of popular fiction give expression to deep-seated cultural and psychological conditions related to our being.

Certain scientific and philosophical discourse that focuses on *cosmology* promotes the multiverse to explain the nature and origins of the universe. In an article in *Scientific American* Alexander Vilenkin and Max Tegmark write: "Intelligent observers exist only in those rare bubbles in which, by pure chance, the constants happen to be just right for life to evolve. The rest of the multiverse remains barren, but no one is there to complain about that."[18] Concepts of an infinite array of different universes with slightly differing laws of physics, constants, constraints, and circumstances help explain how the unlikely existence of a universe, our particular universe, is able to generate sentient organic life, especially humans who are able to reflect on such ideas as the multiverse. There is a spectrum of discourses here. They range from the multiverse as a stimulating thought experiment to mathematical models such as those offered as a means of explaining the paradoxes of quantum physics.[19] I will look into quantum physics and the quantum Internet toward the end of this chapter. Arguments against the multiverse theory include its lack of falsifiability. How could you ever know if other universes exist, especially as these varied universes are beyond our "cosmic horizon"?

I introduced concepts of hidden dimensions through spatial geometry. This is of most interest to my commitment in this book to urban cryptography and secrets, not least as it implicates space and representations of space in computer systems via mappings, transformations, projections, complexity, non-Euclidean models, video gaming, virtual reality, and immersive experiences. My own fascination with the putative experience of these universes resides not just in their characteristics but also in how we supposedly transition from one to the other.

Portals

Where do these parallel universes exist? That's an obvious question for an architect invested in the materiality of the physical environment. The answer could be geographical. These putative worlds could be millions of miles apart. They could exist at different times. Then there's the relationships between them. One may exist as a different state of the other, as solid and liquid are different states of water. Under a psychological account, such universes might exist under different cognitive or bodily conditions, such as waking or dreaming. One world may be subsumed or hidden inside the other as in a steganographic image. One may be secreted in the other's geometry, a scaled-down microscopic version, like a Mandelbrot fractal, or as an image reflected in a mirror. Some worlds may be made of different materials, beyond substance, or even beyond description as substances—invoking distinctions that are after all ineffable.

In this chapter I have taken the view that it is sufficient to accept other worlds as elements of language and storytelling, let alone raw human experience, to regard the possibility of other worlds as a legitimate subject of urban inquiry. At the very least, many elements in fantasy and science fiction stories deal in allegories about a current condition, are vehicles for speculative thought experiments, challenge world views, or simply offer recreation or respite. The multiverse is a way of talking. So, I can reasonably inquire how such narratives deal with transitions between worlds in much the same way that visitors would inquire how they are to get from the Garment District to the Financial District in New York.

If talk about parallel worlds is simply a figurative account of differing points of view, frames of reference, biases, or preconceptions, then the means of transition fits within the usual processes by which we submit to persuasion, challenges to beliefs, correction, revision, openness, and empathy. A psychological "awakening" might suffice as a way of explaining a portal between parallel worlds. As I have already suggested, if talk of the multiverse involves transition between psychological or emotional states then therapies, music, body practices, medications, and mood-altering substances can assist. Stories of multiverses present the process as such, though the means are diverse.

If the parallel worlds belong within metaphysical discourses, such as Plato's Ideas, an afterlife, Nirvana, seven heavens, then there are time-honored methods of transition: contemplation, meditation, death, rebirth, and instruments of conversion, redemption, and salvation.[20]

Cosmological parallel universe theories of the multiverse imply the impossibility of communication between such worlds. By this reading, transitioning between universes in the multiverse is meaningless. By most accounts, the multiverse is a byproduct of a series of mathematical models or speculations about models. The multiverse is a thought experiment.

That limitation doesn't prevent the imagination from entertaining further speculation. Portals between worlds built on whatever foundation are a mainstay of science fiction and fantasy.

Video games, CGI, and immersive VR provide obvious media for exploring the geometries of other-world transitions. In the case of VR the usual means of moving in and out of virtual places includes the head-mounted displays (HMDs) and other physical paraphernalia, and their spectacularization in science fiction. Video games include virtual in-game portals. How players and audiences transition from one universe to the other, or how one universe transforms into another, involve creative computer modeling, code, and the inventive use of geometry and visual effects. My observations here are informed by my recent foray into the video game *Obduction* by Cyan, the original developers of *Myst*. In *Obduction* the player moves between worlds via spherical metal mechanisms that, when activated, take the player to different parts of the alternative worlds. In most fantasy narratives about other worlds the means of getting from one to the other constitutes a puzzle. The means of transmission is a secret. The existence of the other, parallel universes is hidden, as are the means of getting there.

We can also resort to cryptographic allegory to explain such transitions. Like the contents of an encrypted hard drive the parallel reality appears as random noise. We need the private encryption key for the universe's algorithms to transform it into something that makes sense. Another touch point with the cryptographic city is communication. Whether or not a human agent wishes to move from one universe to the other, they may wish to speak with beings who are in it. There is communication in plain text and plain speech, but it's likely that to preserve its hiddenness, the communication is in code.

After Life

Technologies amplify the idea and possibility of a universe beyond everyday experience, often referred to as the *supernatural*. Technologies of communication and transportation help explain, simulate, or suggest what the supernatural is or would be like, especially as people's imaginations extend beyond the limits of current devices, technologies, and infrastructures. In this and other respects technologies provide a pool of metaphors from which to draw in explaining, developing, and even creating the supernatural. Rapid transportation suggests instantaneous appearance and disappearance—levitation, dematerialization, transcendence. Cables and switches, not to mention magnetism and the wireless, point to action at a distance—a staple of the popularized supernatural. After all, early developments in sound recording were accompanied by speculation about contact with the dead. The gramophone provided a means of memorializing "the last words of a dying person" according to its inventor Thomas Edison, from which it was a simple extrapolation to think of communication beyond the grave.[21] Most important, technologies of sound extract and preserve something from the human body that is already evident as a separable "essence," namely our breath, particularly as part of the operations of the voice. The latter already suggests spirit and soul, and technologies of communication amplify this sense of the "soul's apartness."[22]

Media theorist Friedrich Kittler suggests a correspondence between the invention of Morse code and attempts to hear and receive messages from the spirit world: "The invention of the Morse alphabet in 1837 was promptly followed by the tapping specters of spiritistic seances sending their messages from the realm of the dead."[23] An article by media scholar Anthony Enns concurs with a helpful account of this peculiarly Victorian practice: "The main similarity between the spiritualist practice of 'rapping' and the development of electrical telegraphy was that they both involved the translation of messages into codes."[24]

Two young sisters Margaret and Kate Fox famously conducted seances in the United States in the mid-1800s. They toured the UK in 1852, and impressed Arthur Conan Doyle, who explained (via a narrator) the apparent rapping response of a spirit at one of their seances: "I then asked: 'Is this a human being that answers my questions so correctly?' There was no rap. I asked: 'Is it a spirit? If it is, make two raps.' Two sounds were given as soon as the request was made."[25] Enns draws on Conan Doyle's book then reiterates: "The practice of 'rapping' thus involved the use of a binary code that closely resembled the 'dots' and 'dashes' used in telegraphy."[26]

These apparent communications represent a further coded dimension to the cryptographic city, especially the city as a site of hauntings, with otherworldly dimensions, and as a "spiritual" place. "Spirit" has several connotations. There's the "spirit of the times" (Zeitgeist), and the "spirit of place" (genius loci).[27] The "spirit of the city" is also the city of spirits, in other words, the city of mysteries, if not the city of codes.

Nano Worlds

Cities are lessons in physical scale, from over-scaled ambitions such as Bitcoin City's masterplan to the teaming world of microorganisms and intracellular parasites (viruses). Worlds may go undetected as they are at a scale that evades the senses—until an epidemic pauses the life of a city. Developments in biotech also impact the cryptographic city, not least as we consider DNA as a coding medium.

To the popular imagination, DNA is a code system. A strand of DNA is a complex structure consisting of sequences of four molecules connected in pairs in a double helix configuration. These nucleotide molecules are adenine (A), cytosine (C), guanine (G), and thymine (T). The DNA in a human cell is made up of around 3.2×10^9 of these pairs, normally tangled into twenty-three paired strands (chromosomes). ACGT sequences in DNA strands offer the potential to hold vast amounts of binary data, but the processes of manipulating DNA sequences are time consuming and expensive as are the processes of reading these sequences. Biotech researchers consider other methods of exploiting the properties of DNA strands in data storage and encryption.

Genetic engineers use enzymes found in bacteria to cut and join strands of DNA at specific sites along the strands. Strands of DNA can also be reproduced and multiplied by injecting them into bacteria cells. DNA strands can be sorted into different lengths via an apparatus that draws them through a gel toward a positively charged electrode. Genetic engineers can cut DNA strands in a way that creates conditions at the ends of the strand fragments that encourage other strands to connect. Geneticists harvest, sort, filter, and store DNA strands in liquid solutions. The combinations of nanoscale DNA strands can be inspected via electron microscopy.[28]

These particular processes of synthetic biology exploit properties of DNA strands that are independent of how a DNA sequence might actually function in a living organism—to synthesize other complex molecules that perform cellular functions. The artificial processes treat DNA as molecular matter, as bricks, blocks, plates, tubes, and planks to be layered, joined, stacked, and folded. The production process typically involves titrating precise quantities from solutions of each strand with specific properties into a mix that is heated and cooled at some optimal rate to encourage nano shapes to form.

Computer-aided design systems (e.g., caDNAno at cadnano.org) assist with designing 3D shapes, generally following a voxel geometry (i.e., much simplified versions of the kind of voxel art shown in figure 9.6). Data from the CAD system is channeled to robotic titration machines that produce the solutions/suspensions from which these nano shapes can form. There is generally some wastage. One of the goals of the process is to minimize the production of mis-formed shapes and fragments.

Genetic engineers use these techniques to fabricate, isolate, and deploy a class of short DNA strands (twenty or so DNA base pairs) known as *staples* (oligonucleotides). There are techniques for inserting these staples into long DNA strands at specific locations to cause the strand structures to fold. So the double helix of a strand of DNA is effectively a tube able to be bent and folded by the judicious placing of these staples to make shapes. That is *DNA origami*. It is ex-vivo synthetic, controlled lab work, and is independent of operational DNA in living cells.

A helpful article by Swarup Dey and colleagues identifies the range of applications of DNA origami, including, for example, its use in providing masks or templates for printing patterns for nanoscale electronic circuitry. They also mention "light-harvesting antennas and photonic wires with long-range directional energy transfer."[29] Folded DNA strands have a negative charge, a feature that can be exploited to induce rotation. Swarup Dey and colleagues state: "A major goal for DNA nanotechnology is to create molecular machinery and motors that do not just switch between states upon sensing some external change but also are progressively fueled through a closed state path and generate change externally."[30] One popular application is delivering drugs to specific sites in the human body. Dey's article includes an image of a microscopic cube-shaped box with a hinged lid: "DNA origami structures can also serve as containers with docking sites in their interior or within dedicated cavities, protecting the payloads from the environment and the environment from the payloads."[31] This is architecture and nanoscaled urban design of a sort: designed, constrained, responsive to environment, functional, consequential, socially influential, speculative, and risky—but invisible.

DNA origami is predicated on various tropes of hiddenness: nanoscale locked "boxes" made of folded DNA strands to conceal active molecular agents (enzymes, drugs, active DNA material) from their immediate environments where they may be damaged or cause harm. Nano objects are in any case hidden from direct view, only detected via sensing apparatus such as electron microscopes and spectrographs.

Such invisible nanosynthetics engender similar suspicions to encryption, secret messaging, and secrets in general. According to a Reuters fact check site, "Social media users have claimed the presence of lipid nanoparticles in a COVID-19 vaccine means it could contain small robots or computers."[32] The secretive world of cryptography elides with mistrust of difficult science and invisible technologies.

An article "DNA Origami Cryptography for Secure Communication" spells out an interesting method for hiding information in DNA strands that resonates with the manipulation of visible and tangible physical matter.[33] This is not the instantaneous encryption and decryption of binary data in volume, but the secure delivery and preservation of short and discrete blocks of data, such as long-term storage of passcodes and persistent cryptographic keys. The method starts by coding a short message as a grid of dots, a little like braille. Zhang and colleagues illustrate this with the letters H, E, and Y. Each letter is represented on a simple 3×3 array of dots that are either on or off. The bottom row in the array indicates the position of the letter in the message. That is needed as the message eventually appears as a coded "alphabet soup," like random dominoes, and the recipient needs to reorder the letters. These dot patterns are imprinted as identifiable DNA markers on folded DNA nano-"scaffolds," typically flat plates formed of folded DNA strands. The pattern of folding is crucial as the dots will appear in different positions to the original message if the folds are misplaced. The physical DNA mixture is passed from the sender to the receiver: a test tube (or impregnated paper) with the fluid containing the relevant DNA mixed in with other DNA material extraneous to the message but important for the decryption process. Both sender and receiver have access to the "defined DNA origami folding scheme"[34] which is secret data already agreed.

The folding scheme provides the composition of DNA strand chemicals needed to reestablish the scaffold on which the dots are "printed." The recipient essentially mixes chemicals to reestablish the folds (staples) to reform the transmitted DNA nano-scaffold. Centrifuge filters remove surplus short DNA strands, and chemicals are added so that the dots show up on scans. The dot configurations appear on the scaffolds under special fluorescent light (stochastic optical reconstruction microscopy—STORM) and the patterns are interpreted and ordered back into the plain text message: "HEY."

The decryption process in the experiment so described in the Zhang et al. article involves the physical exchange of DNA material; it is painstakingly precise. They describe an experiment in which it took one to two hours to decode each pattern, which I take to mean per character in the message. The process is clearly costly, noisy, and error prone. Nevertheless the researchers claim that the method holds promise of much better security than the Advanced Encryption Standard (AES). They write in summary that their method "uses information-based DNA self-assembly to create physical puzzles, resulting in extraordinarily strong all-around protection of a secret message."[35]

The method recalls the old string cipher method of delivering a secret message described in chapter 2 (figure 2.1), in which a plain text message is coded as a series of knots in a length of string. When the string is unfurled and zigzagged across a correctly proportioned wooden template (scaffold) the knots line up against the letters of the alphabet to reveal the secret message. Both parties have to have a scaffold to the same design. I think the method I have described here serves to defuse any suspicion that genetic engineers are on the verge of using DNA to transmit hidden messages that will impact directly on organic life or human physiology.

Obvious possibilities for nanotechnology in the city reside with innovations in medicine, agriculture, the manufacture of new materials, and other outcomes of bioengineering. The design of hinged intracellular containers and mechanisms recalls my discussion in the introduction of the affordances of ordinary hinged doorways. I'm prepared to think that the physicality of nanotech methods and artifacts entitles architects and urbanists to include the field within the orbit of the city. That these technologies might also be used as tools for encryption draws them further into the cryptographic city. To build on the theme of different scaled worlds, if cryptography is relevant at this intracellular nanoscale, then it also applies at the interstellar scale.

Interstellar Cryptography

Other worlds may exist simply as other places spatially distant from us across interplanetary and interstellar distances. Such expanses are sufficient to facilitate physical transportation from one world to another but at exorbitant cost. Communication with those worlds provides similar challenges.

We might presume that the aim of cryptography is to make a signal as obscure as possible to anyone but the designated recipient. *Anticryptography* is a loose term to designate a type of cryptographic message that is legible to someone who has *no* knowledge of the plain text language from which the message derives. Nor do they have access to the method of encryption, or anything like an encryption or decryption key. Nor is the message meant to be secret. The message is coded in such a way that a codebreaker from any language group, culture, or context has a good chance of decoding it.

Why would anyone want to create such a code that is designed to be easy to decipher? It is for sending messages to extraterrestrials and making a good guess at how alien intelligences might already try to communicate with us: "Probably the one overriding principle of the outer-spacelings will have been to make their message as clear as possible. It will be coded, but in a code designed for clarity and not for obscurity—a kind of cryptography in reverse, . . . an anticryptography."[36]

The coding system called *Lincos*, from "lingua cosmica," was invented by the mathematician Hans Freudenthal (1905–1990) who describes the system in his book *Lincos: Design of a Language for Cosmic Intercourse.*[37] The book is dryer than the title suggests. Figure 13.1 shows a sample dialogue in a symbolic language derived from lincos.[38] Freudenthal was invited to develop his code by the British Interplanetary Society in the 1950s.[39]

Lincos qualifies as a cryptographic system rather than a *language*. The sender assumes that the potential recipient has no knowledge of the system, or even that a signal they receive is a coded message. The recipient has to be able to detect that the transmitted signal is likely to contain a message and is not random noise or an extraneous signal generated by natural phenomena. Freudenthal assumes the medium for a Lincos message is electromagnetic pulses. Whatever the medium, the message has to repeat to increase the chances that it will be picked up by someone or something that can read it. The pattern of the repetition may also offer a clue that the signal is a message from intelligent life. According to Kahn, "Lincos would have to be taught to the creatures of outer space before it could be used as a medium of communication, and Freudenthal proposed to do this by transmitting the statements of Lincos, which he hoped would be relatively self-evident, over and over again until the recipients catch on to their meaning."[40]

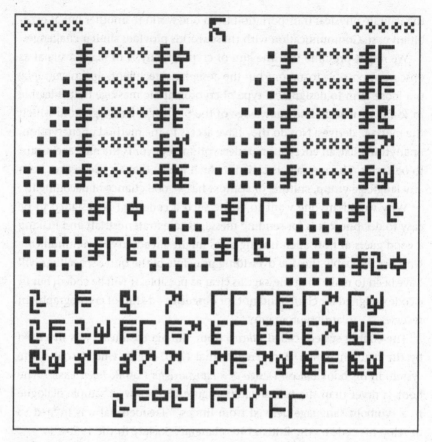

Figure 13.1
The first page of code transmitted from the Evpatoria radar telescope in Ukraine on May 24, 1999. Inspired by Freudenthal's Lincos system it begins by defining arithmetic operations, the elements, units of measurement, and our solar system. The last four lines of the first page show the first ten prime numbers, ending with the highest prime known at the time. Each symbol is 5x7 pixels and designed so that if a single pixel in a symbol is misplaced in transit, then the symbol will not be misread. For an explanation of each symbol see Daniel Oberhaus, *Extraterrestrial Languages* (Cambridge, MA: MIT Press, 2019), 179–191. *Source:* MIT Press, used with permission.

The aim of delivering messages in such a way that the means of interpretation and translation are part of the code recalls my argument in chapter 3, and the idea that urban artifacts tell us how they are to be used without explicit instruction but in the way they are designed. The process is even more difficult in the case of communicating with putative aliens as we do not know if they share the same relationships with their environment as we do with ours.

The motivation for interstellar signaling came from speculations about life on Mars. Once detected by telescopes, observation of the "canals" on Mars encouraged people to invent means of communicating with Martian life, with the best chance of contact occurring when the orbits of Mars and Earth brought them closest. The polymath Francis Galton (1822–1911) identified the need for such communication as early as the 1890s.[41] He proposed "scintillations of light" [42] as the likely means for Martian intelligence to reach out to us.

Others have proposed less-general means of communication. Karl Friedrich Gauss (1777–1855) had proposed planting a massive forest across Siberia in a configuration of triangle and squares that would leave an extraterrestrial in no doubt that we earthlings know Pythagoras' theorem. Sprawling cityscapes and their illumination would now provide similar signs of earth life. Unsurprisingly, attempts at a celestial code center on the assumed universality of mathematics. A second component is the context of the signal. A potential receiver might expect the communication to establish and identify the location of the sender, which could be correlated with other observations about our own planet's location in the universe. The astronomer Frank D. Drake has added the likelihood that any extraterrestrial codebreaking would involve teams with different expertise, as it would on Earth. Kahn quotes Drake who writes as if the communication had already succeeded: "In preparing the message, an attempt was made to place it at a level of difficulty such that a group of high-quality terrestrial scientists of many disciplines could interpret the message in a time less than a day. Any easier message would mean that we are not sending as much information as possible over the transmission facilities, and any harder might result in a failure to communicate."[43]

I doubt many earthlings would have the time or the inclination now to learn precisely how Lincos operates. Kahn provides a useful summary: "He began his program by sending a series of messages to teach the terms 'plus' and 'equals.' His first message might be beep beep beep beep bloop beep beep tweet beep beep beep beep beep beep. Next he might send beep beep bloop beep tweet beep beep beep. After sending enough of these for the outerspacelings to catch on to the idea that bloop is 'plus' and tweet is 'equals,' he might transmit a message with a new signal, like beep beep beep blip beep tweet beep beep. Soon the spacelings would realize that blip means 'minus.' Similarly, Freudenthal would build up an entire mathematical vocabulary."[44]

As I have already suggested, the idea that a cryptographic code might be designed to be broken is interesting from the perspective of the built environment. Anticryptography is a version of the challenge we encounter all the time in communication. It is the challenge of learning one's first language as an infant. We don't have to memorize a code before we can speak and understand. We learn language as we use it, from mimicry, context, feedback, and a certain inbuilt propensity to communicate with highly variable combinations of sounds. As explored in chapter 3 on urban affordances the process is the same with the built environment. Devices, machines, buildings, streets, furniture, and parks convey the means of their understanding, or they could if designed accordingly.

Simultaneous States

Arguments for the existence of the multiverse find support from theories within quantum physics. Related ideas about the quantum Internet (QI) impact on operations within cryptography and add further layers to communications infrastructures in the city.[45]

I will start with lasers. Laser technology is ubiquitous in the city, not least in LiDAR for measuring, surveying, and scanning (figures 5.1 and 6.1). A laser beam is a concentrated beam of coherent light within a narrow color band (i.e., frequency range). An experimenter can point a laser beam at a piece of card that has two slits cut close together into it.[46] Some of the laser light will pass through the slits. A surface on the opposite side of the slits to the laser light source will show an interference pattern of light and dark bands. That is because light beams propagate in a wave pattern. The light waves are refracted by the two slits and spread out creating two wave patterns. The interference pattern appearing on the surface is brightest where the light waves reinforce one another, and will be darkest where they cancel each other. So light seems to behave as waves do.

Light is also observable as individual particles, photons. Experiments that track the passage of low-intensity laser light with countable numbers of photons shows that individual photons do in fact exhibit this wave characteristic when observed en masse. A sensor can count the photons and record their positions as they arrive on the surface in the two-slit experiment.

Photons are discrete particles and don't exhibit degrees of luminance. Nor do they reinforce or cancel each other like waves. The light patches on the screen are simply where the photons land in higher numbers. The dark patches have no photons. If one of the slits is covered over, then the photons entering the open slit will fall on the screen without any interference banding.

This two-slit experiment highlights one of the paradoxes in the behavior of light or the movement of any subatomic particle. It is as if photons passing through one slit take account of the photons passing through the other slit so they agree, concur, or conspire where they will fall if the light was moving as waves. The results of these experiments imply that under the right conditions photons from the same source coordinate with each other. To avoid the language of agency, that linkage is termed *quantum entanglement*. The description I've just presented is standard fare in physics textbooks.[47] The two-slit experiment establishes that photons exhibit quantum entanglement. The behavior of one particle depends on what happens to any other particle emitted from the same source, even though they are observed at different times and places. Hardware engineers and scientists have recruited this phenomenon to propose and develop a radically new class of computers. These are *quantum computers* that exploit the properties of quantum entanglement.

Quantum computers are extremely fast computational devices that exploit quantum phenomena. They exist in labs and theoretical models at present, but promise to outperform current, everyday supercomputers by many orders of magnitude. As well as the two-slit interference phenomenon, in quantum computing the characteristic of the photons of interest is their *spin*. Subatomic particles such as electrons, protons, and photons spin in two directions, usually described as "up" or "down." It is tempting to think of a ball spinning rapidly, stopping and changing direction, oscillating from one to the other or rotating about two axes. Were you to freeze a few frames in a video of the spinning ball you would catch it in one direction or the other, or catch it between states. This is the wrong model, though. Spinning electrons do so in two directions simultaneously and with no intermediate states where they transition from one to the other. Specialized sensors can detect and measure the spin direction of a photon. Until the spin of a photon is measured, it is in both states (up and down) at once, simultaneously. Laboratory experiments can exploit the property of quantum entanglement to coordinate the measurement of the spin states of two photons that have come from the same coherent light source, even though the photons have moved far apart.

The basic unit of arithmetic in a computer is the bit, an electronic component in a microchip that can be in one of two states.[48] It is either on or off, 1 or 0. Microchips contain circuitry that stores and manipulates strings of bits at billions of times per second. Unlike a binary computer, the basic unit in a quantum computer is the *qubit*, a cute name that indicates a unit that stores a value of either 1 or 0 or both 1 *and* 0. The condition of being in two states at once is the nub of the quantum mechanical property of subatomic particles. This is the middle state of a qubit. In the case of a spinning electron it stays in that dual state until someone interferes to measure it. Then the spin is either up or down according to some probability.

A quantum computing microchip has circuits that exploit arrays of supercooled subatomic particles spinning in this controlled quantum state, but without extraneous electromagnetic interference or measurement, until monitored to record their states at that moment. The idea of "recording" doesn't do justice to the mechanism here, which under quantum logic is to both measure and create. The state of an array of qubits is computationally interesting. Fifty or so qubits connected each to the other provides an instantaneous parallel processing unit. The size of the array doesn't impede the quantum calculation, and the computational power increases exponentially with the addition of extra qubit processing capability.

There are parallels here with chemical processes in which molecules interact without iteration as required in normal computer simulations. So biotech is one of the key areas motivating the development of quantum computing. That and security, fintech, and the possibilities of a QI.

Exploiting the procedure of the two-slits experiment, it is possible to emit two photons from a lab device made up of a laser and refracting prisms. The photons will have the same initial quantum state. As they come from the same coherent light source, they will exhibit quantum entanglement. The two photons remain entangled as long as nothing interferes with them.

They will continue in this quantum state as they pass through space or a fiber optic cable. Experiments show that a pair of photons can retain quantum entanglement even though they travel away from each other—up to 100 kilometers. Even though the spin of the photons is indeterminate, were someone to measure the spin of one of them it would be either up or down with a probability of 50 percent. If someone were positioned with the other photon, then they would get the same reading as the photons are still entangled at a distance. The senders of a message cannot change the spin of a photon, but once they have detected the spin they can be sure (to a known probability) that the recipient has the same reading. That has implications for securing communications. The phenomenon does not permit message sending "faster than the speed of light," though that is an alluring imaginative scenario.

To take a reading is to erase the photons' quantum entanglement. Once the two photons become disentangled then the ability to predict the state of one photon by knowing the state of other is broken. So a quantum network made up of photons coursing through a network requires a stream or steady pulse of photons. Physicists have invented "quantum repeaters" that enable these networks to be extended. One of the main advantages of widely distributed networks that support quantum communications is in connecting quantum computers together so that they can share processing of large scale problems (e.g., in bioengineering).

The security advantage of the quantum Internet is that the state of a qubit cannot be copied. The quantum entanglement between the photons would be severed were someone to take a reading of one of the photons while it is in transit. That is, interference would be detected at either end of the communication channel.

A further advantage of quantum communication is accurate synchronization between equipment on earth and in satellites. The GPS (global positioning system) relies on accurately synchronized clocks on the satellite transmitters and the earthbound GPS receiver to determine distance and hence triangulate the location of the receiver, your GPS-enabled smartphone. An article in *Nature* predicts geolocation that deploys quantum communication with accuracies in the order of millimeters, improving on the one meter accuracy of current GPS.[49]

The quantum Internet suggests a new layer to urban communications infrastructures. Discussion about the QI involves network nodes positioned across cities. Though the QI will not enable communication without radio waves or other media, it has the potential to enhance secure communications: "Entanglement can't directly transfer information, because that would mean data is traveling faster than light. But entangled particles can be used to create secret 'keys' that enable extraordinarily secure communication."[50]

Quantum computers also afford risks to the security of conventional communications. They can potentially remove the need to iterate through the huge numbers of combinations required to break a code, reconstruct an original source document from a hash string, or derive the key used to encrypt a file. Some cryptography experts suggest quantum computers could thereby break the blockchain underlying bitcoin and other cryptocurrencies. Without some remedial action, the wealth tied up in cryptocurrencies would disappear.

These risks have spawned an industry of *postquantum* encryption algorithms designed to thwart the capabilities of these computers of the future. Cryptographers are also developing encryption methods based on quantum computing that will be even more robust than current methods. Keith Martin's book *Cryptography: The Key to Digital Security, How It Works, and Why It Matters* provides a helpful summary of some of the issues in a section ominously titled "Weapons of mass decryption."[51] Competition to develop practical quantum computing internationally and commercially is intense and the stakes in leading in this technology are high. China has already launched a "quantum satellite" as reported in *New Scientist*.[52]

Forking Paths

Explanations of quantum physics involve mathematical formulations incorporating concepts to which I've already alluded in this book, such as hidden variables, coefficients, wave functions, signal processing, spectrographs, probability, and navigating through spaces of potential solutions. I won't here attempt an integrated mathematical theory connecting quantum physics with cryptography. But outside of the mathematics, the field has long prompted physicists and philosophers to propose interpretations about what quantum phenomena mean for our understanding of the world—Neils Bohr (1885–1962) and Werner Heisenberg (1901–1976) among them.[53]

The most interesting theory from the standpoint of the discussion in this chapter pertains to the *many worlds hypothesis* advanced by the physicist Hugh Everett (1930–1982) in a seminal 1957 article. In chapter 5, I discussed the labyrinth to introduce a series of concepts that pervade the cryptographic city. Moving along a branching path, or through a city, involves decision points, junctions at which the traveler continues the journey along one of several paths, even looping back at times. The idea of forked timelines provides a means of accounting for the paradoxical observations evident in quantum physics. Everett's innovative contribution is grounded in mathematics, though he explains the theory in everyday terms as a branching process. He maintains that the observer of the state of a spinning particle is in multiple states: "With each succeeding observation (or interaction), the observer state 'branches' into a number of different states. Each branch represents a different outcome of the measurement and the corresponding eigenstate for the object-system state. All branches exist simultaneously in the superposition after any given sequence of observations."[54] By "eigenstate" he means the definite measured value of a particle state such as the orientation of its spin (up or down). "Superposition" is the combination of two quantum states (both up and down).

This formulation is analogous to traversing each path leading from a junction in a maze simultaneously. The mathematical formulation by which Everett derives his assertion about branching states puts the focus on the timeline of the human observer: "The 'trajectory' of the memory configuration of an observer performing a sequence of measurements is thus not a linear sequence of memory configurations, but a branching tree, with all possible outcomes existing simultaneously in a final superposition with various coefficients in the mathematical model."[55] Though quantum physics is counter-intuitive in many respects, Everett claims that it accords with experience, at least in the laboratory: "The theory based on pure wave mechanics is a conceptually simple, causal theory, which gives predictions in accord with experience."[56]

Others extend quantum theory beyond laboratory experiments.[57] I would note that Everett's many worlds theory does resonate with everyday human experience, in particular the human propensity to speculate about alternative courses of events, to breathe a sigh of relief when things work out, or regret when they don't. The idea of branching timelines predates quantum physics, though the latter has injected a semi-scientific aspect to such narratives.

Two films in the 1990s give expression to branching states existing somehow in parallel. *Sliding Door* (dir. John Madden, 1998) shows what happened to her timeline when a woman was either on time or too late for the closing door of a subway car. *Run Lola Run* (dir. Tom Tykwer, 1998) delivers a series of narratives and subnarratives in a twenty-minute dash by a woman to get a large sum of money that would save her boyfriend's life. The return and replay of events are marked by an extended primal scream from Lola at crucial moments of regret. The connection between alternative timelines and quantum physics is explicit in the mini-series *DEVS* (2020) directed by Alex Garland. It weaves alternate scenarios into a high-tech narrative about a quantum computer. At one point in an episode one scientist dares the other to test his faith in the multiverse by letting go of the guardrail while balancing on the edge of a dam wall. Whether he stays safe or falls depends on the path adopted by this nihilistic drama. Our awareness of risk, the excitement of not knowing an outcome, and the impulse to gamble participate in this multiverse sensibility. That the young scientist both falls and returns to safety in branching timelines is not verified, except we can see such multiple outcomes in fiction, film, and our imaginations. Branched narratives are of necessity selective to create a compelling story, though sometimes the offer of alternative timelines become part of the narrative.[58]

The experiential aspects of quantum physics yield interpretations that appeal to scholars invested in a phenomenological and pragmatic understanding of the world.[59] Further support for this pragmatism comes from the quantum pioneers themselves. Neils Bohr noted the way measuring instruments are complicit in the phenomena observed. He saw that for observations and calculations about subatomic particles "no sharp distinction can be made between the behavior of the objects themselves and their interaction with the measuring instruments."[60] Heisenberg (1901–1976) inflected that insight with his assertion that "what we observe is not nature in itself but nature exposed to our method of questioning."[61]

Everett's theory about parallel worlds relates to multiple timelines in the case of laboratory observations. If quantum states are under constant "observation" by animate and inanimate matter, and via sentient and insensible agency then the proliferation of many worlds extends infinitely.

Drawing on such sources, theoretical physicist Karen Barad in her book *Meeting the Universe Halfway* develops a phenomenological account of quantum physics, inflected by her readings in feminist critical theory and posthumanism: "Phenomena are not the mere result of human laboratory contrivances or human concepts. Phenomena are specific material performances of the world."[62] In other words, "meaning is an ongoing performance of the world."[63] She mentions quantum cryptography in passing as a potential source in her argument. I think that such insights into the world as performance resonate with ideas that place has agency, is its own code—that elements within settings such as cities communicate and deliver not only messages, but also the means of their decryption, which is to say their interpretation. There's a pleasing confluence here, a conflation of "observer and observed, knower and known"[64] or "words, knowers and things."[65]

In this chapter I expanded the hidden dimensions of space and geometry to spaces as products of the imagination. Cities so defined are cryptographic in that they are hidden, require some investigation to uncover, and require some process of communication to transition between them. How do we enter these alternative universes? Here, the allegory of cryptography overlaps or underpins urban imaginaries. These themes come to prominence when we think of the alternative worlds suggested by digital manipulation, virtual reality, cybercities, and many worlds. Like the contents of an encrypted hard drive the alternative reality appears as random noise, until the private encryption key makes sense of it. That is via a portal requiring a code, a key, a secret knowledge. We need the private encryption key for the universe's algorithms to transform the apparently unknowable into something that makes sense. Note the use of visual noise on the computer screens in *DEVS*. In that fiction, images of the past and future emerge from a fog of random pixels as the quantum processing improves.

Speculations about extraterrestrial communications emphasize that the means of decoding are in the code, and developments in quantum physics expand further the human fascination with parallel universes and secret realities, each of which impinges on encryption methods and the cryptographic city. That leads me to consider again that the means of decoding the city are already within the city, though specialized technologies and algorithms provide innovative means of supporting urban affordances.

14 Closure: The Urban Cryptanalyst

> If someone figured out my secret 4-digital PIN then I would really be in trouble. They'd have my bank PIN, could access the door to my office, and get into my smartphone. Worst of all, they would know the year of my birth!
> —Adapted from https://upjoke.com/code-jokes

Codes are ubiquitous. I hope that I have been able to demonstrate that cryptography is a major element in the contours of any city, delineated by technologies, instruments, and processes for hiding and revealing information, messages, things, spaces, places, and people. Cryptographic devices are built into the forms, organizational arrangements, framings, myths, and metaphors of the city. The tenets of cryptography align with theories about cities. In this book I have expanded on the cultural aspects of cryptography and the city. Cryptography is as old as secrets. Society could not function without secrets, and that sustains our curiosity. Whether or not we invent, code, and maintain encrypted systems, we urban inhabitants are unintentional codebreakers constantly seeking out meanings and how to interact with the city, which is to say to interpret our world.

I grouped the discussion in this book under four headings. The first made the case for an urban cryptology, with an emphasis on cultures of writing. The second introduced the combinatorial aspects of making, writing, and reading the city. The third delved into algorithms and calculations. The fourth examined some of the outer limits of cryptographic operations.

The Case for an Urban Cryptology

Some architects have drawn on the mystique of esoteric symbols and numbers for inspiration and catered to the requirements of secret societies. I identified key affordances of cryptography: to reveal and conceal, combine and enumerate, originate and navigate, activate and transact, engender risk and trust. My purpose here was to collate the most obvious cases where architecture and the city are inscribed with coded messages in stone, concrete, and steel before pursuing the less obvious connections with cities in the rest of the book.

The history of writing and print is relevant not least as it implicates encryption as a process of rearranging letters in movable type. It is also a touchpoint for architecture and urbanism through the innovative cryptographic apparatus (the cipher wheel) of Leon Battista Alberti. Alberti's cipher wheel and his influence in architecture and urbanism entitles architecture to claim a pivotal place in the history and theory of cryptography, and hence as custodian of the cryptographic city. Understood through sign systems and semiotics, appropriately designed places provide the means of their own decryption. Place is the code, a proposition supported by concepts of *affordance* from environmental psychology and media studies.

Urban Combinatorics

Cities are places of combination and recombination. The theme of combinatorial complexity led me to review the nature of riddles, jokes, and puzzles to furnish explanations of city affordances. To view the city through the lens of cryptography is to see the city as a spatial puzzle. Labyrinths provide a further leitmotif and metaphor for the cryptographic city's infrastructural complexity. Digital platforms and programs that find routes and navigate through cities and problem spaces are commonly tested with networks shaped like branching labyrinths. Much digital simulation of city functions and spaces invariably involves disguise, dissimulation, a kind of play function.

Cryptotechnics

Cryptocurrencies offer high-profile, disruptive, and far-reaching examples of applied cryptography. Attention to cryptocurrency and the blockchain emphasizes the consequences of this investigation into the cryptographic city, not least the central role of transactions involving money. A city is also an array of locks, keys, vaults, hidden spaces, security doors, cameras, contactless sensors, keypads, passcodes, biometric IDs, and face recognition—fixed and mobile. Under the blockchain metaphor, cities reveal themselves as hyper-encrypted and hashed (a kind of recombination), and the city depends on that.

I examined encryption methods that pertain to images. The city as a collage suggests time-bound and dynamic interactions between erased, hidden, and fragmented layers. Steganography as a branch of cryptography operationalizes certain aspects of hidden layering. This introduced the concept of the hash string, ubiquitous in encryption methods, as a kind of signature. I examined the methods behind image compression, authentication, the specialized procedures of hashing, hashes of hashes, chained hash strings, and proof of work, as the bases of blockchain and other encryption techniques in the cryptographic city.

Cryptography at the Limit

Cryptography belongs within tactics of obfuscation and espionage—urban tactics deployed by state instruments, corporations, activists, and day-to-day creators and users of information. Obfuscation enters the field as a major element in cyberattack and cyber defense strategies within cities.

I examined the nature of algorithms and hidden variables as vehicles for covert messaging. The term *hidden dimension* is architecturally alluring as it implies there are spatial parameters that extend beyond our usual experiences of a place or are accessible only via specialized inquiry. The exposure and manipulation of a city's hidden dimensions is emblematic of social and digital processes in the cryptographic city. The challenges offered by cryptographic communications at the nano (DNA) and interstellar scales emphasize the scope of cryptography's claims on the physical world. Putative extraterrestrial communication attempts to provide the means of decoding in the coding mechanisms themselves, offering instrumentalized support for the idea of urban affordances. Developments in quantum physics expand further the human enchantment with alternative universes and secret realities: the cryptographic city as experienced and imagined.

The Accidental Cryptographer

My arguments presented four main claims about urban cryptography. First, was the obvious claim that cryptography provides a way to address and solve some of the challenges of data security and information flows in the city. I have left it to others to advocate for vigilance in adopting appropriate security practices. The second claim was that people, objects, and information hide in cities. By inspecting the city through the frame of cryptography, we understand better the hidden aspects of city life and form. In this case cryptography provides a pretext for talking about urban hiddenness. Third was the claim that cryptographic processes mirror and inform what happens in the design, management, and use of city elements and spaces not least through the idea of the city as a process of combination and recombination. Fourth, the challenge of cryptography in the city is less a problem of coding and hiding as decoding and exposing. *Accessibility* is the watchword of contemporary urbanism: laying bare and revealing rather than hiding things away.

I have emphasized cryptography in the conceptual framing of the city. Urban researchers have conducted similar exercises through diverse framings: the city through the imagination, utopia, night life, walking, skateboarding, seasons, light, color, the transhuman, wellbeing, disadvantage, division, contest, efficiency, prosperity, and smartness. Each framing sheds light on different facets of the city—and a city is surely multifaceted. Each also occludes other aspects of the city not accommodated in any particular framing.

What does a conceptual framing of the city as *cryptographic* reveal and conceal? It reveals the city as a site of secret societies, pervaded by the remnants of a less-familiar past, of affordances encoded into artifacts, a place of writing, forensics, puzzles and combinations, labyrinths, underground places, disguises, ubiquitous covert communications and transactions, barter and unregulated commerce, clandestine imagery, espionage, algorithms, hidden dimensions, and parallel existences. As indicated by the naming of the *cryptographic* city it exposes a range of facets of city living, but also admits the inevitability of occlusion, of concealment. Inevitably, this framing emphasizes calculation and technological mediations, while casting the many other varied aspects of urban experience into the shadows, at least for a time.

Notes

Introduction

1. I explored the role of acoustics signals in defining spaces in Coyne, *The Tuning of Place.*

2. Permissionjunkie, "What Secret Codes Are All around Us That Only People 'In the Know' Recognize?"

3. Baraniuk, "The Secret Codes You're Not Meant to Know."

4. This is the practice of runners in the "hare and hounds" activity of the global Hash House Harriers association, whose members adopt a particular set of waymarking signals. See https://www.hashhouseharriers.com.

5. Lessig, *Code and Other Laws of Cyberspace.*

6. Singh, *The Code Book,* 10–11.

7. Brown, *The Lost Symbol,* 274.

8. Kahn, *The Codebreakers,* "A Few Words," Kindle.

9. Sebeok, *Signs.*

10. I have expanded on the theme of interpretation in other texts. See Snodgrass and Coyne, *Interpretation in Architecture,* drawing on the theories of Gadamer, *Truth and Method.*

11. E.g., Freedom of Information Act (FOIA) requests.

12. For the variety of research on the smart city, see Aurigi, "No Need to Fix"; Cecco, "Google Affiliate Sidewalk Labs Abruptly Abandons Toronto Smart City Project"; de Lange, "The Smart City You Love to Hate"; Figueiredo, Krishnamurthy, and Schroeder, *Architecture and the Smart City*; Greenfield and Nurri, *The City Is Here for You to Use*; Hnilica, "The Metaphor of the City as a Thinking Machine"; McDermott, "New Sensorial Vehicles"; Penn and Al Sayed, "Spatial Information Models as the

Backbone of Smart Infrastructure"; van Ditmar, "The IdIoT in the Smart Home"; Willis and Aurigi, *Digital and Smart Cities.*

13. Degen and Rose, *The New Urban Aesthetic,* 77.

14. Berger and Iniewski, *Smart Grid Applications, Communications, and Security.*

15. Mattern, *A City Is Not a Computer,* 16.

16. Degen and Rose, *The New Urban Aesthetic,* 71.

17. Burgess, "What Is the Internet of Things? WIRED Explains."

18. McDermott, "New Sensorial Vehicles"; Yoo, Eun, and Jung, "Drone Delivery."

19. Alibasic et al., "Cybersecurity for Smart Cities."

20. Alibasic et al., "Cybersecurity for Smart Cities."

21. Gold and Shaw, "What Is Edge Computing and Why Does It Matter?"; Gusev and Dustdar, "Going Back to the Roots"; Hamilton, "What Is Edge Computing."

22. Cerrudo, "An Emerging US (and World) Threat," 9.

23. Cerrudo, 14.

24. Cerrudo, 14.

25. Graham, "Software-Sorted Geographies," 564. Also see Graham and Marvin, *Splintering Urbanism.*

26. Mattern, *A City Is Not a Computer,* 16.

27. For an account of how narratives about economics inform citizens' understandings of innovations such as cryptocurrency, see Shiller, "The Bitcoin Narratives."

28. Lipman, Sugarman, and Cushman, *Teleports and the Intelligent City.*

29. See Joss et al., "The Smart City as Global Discourse."

30. Lyons, "Getting Smart about Urban Mobility"; Deakin and Reid, "Smart Cities."

31. Mattern, *A City Is Not a Computer,* 13.

32. Jung et al., *Man and His Symbols,* 236.

33. Zuboff, *The Age of Surveillance Capitalism,* 95.

34. Martin, *Cryptography.*

35. Martin, 7.

36. Ellison, *A Material History of Early Modern English Cryptography Manuals,* 8.

37. Ellison, 8.

38. Ellison, 8.

39. Ellison, 8.

40. Ellison, 8.

41. Ellison, 9.

42. Ellison, 8.

43. Hodges, *Alan Turing*.

44. Harari, *Sapiens*.

45. That's an interpretive enterprise, a hermeneutical process. See Gadamer, *Truth and Method*; Snodgrass and Coyne, *Interpretation in Architecture*.

46. Prominent Trump advocate R. W. Giuliani promoted conspiracy theories about the 1920 U.S. presidential election via his YouTube channel; see, for example, Giuliani, "Election Theft of the Century," since removed for violating YouTube Community Guidelines.

47. Gibson, *The Ecological Approach to Visual Perception*, 127; Gibson, *The Senses Considered as Perceptual Systems*. Note the broad reach of affordance theory to include "media affordances." See Rice et al., "Organizational Media Affordances."

48. Norman, *The Design of Everyday Things*. For the application of affordance theory to care facilities, see Topo, Kotilainen, and Eloniemi-Sulkava, "Affordances of the Care Environment for People with Dementia." Also see the application of Gibson's affordance theory to landscape architecture: Heft, "Affordances and the Perception of Landscape."

49. Gibson, *The Ecological Approach to Visual Perception*, 127–128.

50. Anon., "Affordances."

51. Mumford, *The Culture of Cities*, 3.

52. See my previous book on the pragmatic philosopher Charles Sanders Peirce: Coyne, *Peirce for Architects*.

Chapter 1

1. Lindon, *The Alchemy Reader*.

2. Hyde, *Trickster Makes This World*.

3. Underwood, *Numerology, or, What Pythagoras Wrought*, 2.

4. Scholem, *Major Trends in Jewish Mysticism*, 255.

5. Gaffarel, *Unheard-of Curiosities Concerning the Talismanical Sculpture of the Persians*.

6. Lebeuf, "The Alphabet and the Sky."

7. Lebeuf, 320.

8. Kamczycki, "The Kabbalistic Alphabet of Libeskind."

9. Oliver, *The Pythagorean Triangle*.

10. Based on the novel by Pullman, *Northern Lights*.

11. Brown, *The Lost Symbol*.

12. Curl, *The Art and Architecture of Freemasonry*. *Life* magazine from 1956 provides a set of colored images that highlight the spectacle of some of the Masonic rituals at that time. Anon., "Busy Brotherly World of Freemasonry."

13. For Jacques Derrida, that kind of reasoning would be "intertextual." See Coyne, *Derrida for Architects*.

14. Bogdan and Snoek, *Handbook of Freemasonry*.

15. Prescott, "The Old Charges," 33.

16. Stevenson, "The Origins of Freemasonry," 50.

17. Stevenson, 50.

18. Knoop, Jones, and Hamer, *The Early Masonic Catechisms*, 22.

19. Oliver, *The Pythagorean Triangle*.

20. Dickie, *The Craft*.

21. Dickie, "What the Freemasons Taught the World about the Power of Secrecy."

22. Dickie.

23. Knoop, Jones, and Hamer, *The Early Masonic Catechisms*, 197.

24. Curl, "Freemasonry and Architecture," 557.

25. Curl, 586.

26. Watkin, "Freemasonry and Sir John Soane," 402.

27. Watkin, 402.

28. Furjan, *Glorious Visions*.

29. Sir John Soane's Museum London, https://www.soane.org.

30. Furjan, *Glorious Visions*, 172.

31. Furjan, 172.

32. Furjan, 7.

33. Furjan, 61.

34. Furjan, 112. A translucent LiDAR scan of the building by scanlabprojects.co.uk further reveals its visual attributes. See Anon., "A New Way to Experience Sir John Soane's Museum."

35. Knoop, Jones, and Hamer, *The Early Masonic Catechisms*, 216.

36. Åhlén, "Pigpen Cipher (Decoder, Translator, History)."

37. Lakoff and Johnson, *Metaphors We Live By*. Lakoff and Johnson characterize metaphors in terms of *entailments* rather than *affordances*. I take these two terms to mean the same thing in this context.

38. Evans, "Figures, Doors and Passages," 90.

39. Knuth, "Two Thousand Years of Combinatorics." Also see Knuth, *The Art of Computer Programming*.

40. Singh, *The Code Book*, 30.

41. De Hoyos, "Masonic Rites and Systems," 355.

42. De Hoyos, 355–356.

43. Aristotle, *The Ethics of Aristotle*, 101.

44. Palladio, *The Four Books of Architecture*, 1.

45. Snoek, "Masonic Rituals of Initiation," 323.

46. Matthews, "Mazes and Labyrinths."

47. Thiel, "Processional Architecture."

48. Jones, *Design Methods*. Also see Fraser, "Design Research in Architecture, Revisited."

49. Vidler, *The Writing of the Walls*, 91.

50. We explored the theme of origins in architecture in Snodgrass and Coyne, *Interpretation in Architecture*.

51. Vidler, *The Writing of the Walls*.

52. Odgers, Samuel, and Sharr, *Primitive*.

53. Evans, "Figures, Doors and Passages," 90.

54. Knoop, Jones, and Hamer, *The Early Masonic Catechisms*, 178.

55. On the subject of thresholds, see Coyne, *Cornucopia Limited*.

56. Dickie, *The Craft*.

57. Gorleé, *Wittgenstein's Secret Diaries*.

58. The house was in the austere modernist style. He worked on the project with a trained architect, though by all accounts the design followed Wittgenstein's specifications and he obsessed over the details. See Sennett, *The Craftsman*; Wittgenstein, *Philosophical Investigations*.

59. Scholars have drawn similarities, and differences, with Heidegger's simple hut retreat in the German Black Forest. See Sharr, *Heidegger's Hut*.

60. Wittgenstein had intimate relations with fellow student Francis Skinner to whom he dictated some of his thoughts in philosophy and with whom he holidayed and cohabited for a period, a relationship made public only many years later. Gibson and O'Mahony, *Ludwig Wittgenstein*.

61. For an entry on "Mysticism," see Glock, *A Wittgenstein Dictionary*.

62. Weston, "The Lantern and the Glass."

63. Wittgenstein, *Tractatus*, 90.

64. Wittgenstein, *Philosophical Investigations*.

65. Wittgenstein, *Tractatus*, 25.

66. Wittgenstein, 87.

67. Wittgenstein, 54.

Chapter 2

1. The quote continues by referring to an epidemic in the city: "The pestilence doth still increase amongst us we shall not be able to hold out the siege without fresh and speedy supply." Wilkins, *Mercury*, 4.

2. Wilkins, 69.

3. Ellison, *A Material History*.

4. Ellison, 6.

5. The *Handbook of Semiotics* lists Wilkins among the progenitors of modern semiotics. See Nöth, *Handbook of Semiotics*.

6. Mitchell, *Placing Words*.

7. Mitchell, 11. *Intertextual* is a term used by Jacques Derrida as a tactic in argumentation where the reader is led along a circuit of interrelated terms. I elaborate on this approach in Coyne, *Derrida for Architects*.

8. Mitchell, 16.

9. For my own thinking on the application of Derrida's ideas about language and architecture, see Coyne, *Derrida for Architects*.

10. Mitchell, *Placing Words*, 17.

11. Mitchell, 17.

12. Mattern, *Code and Clay, Data and Dirt*, xi.

13. Mattern, xi.

14. Heyworth, "Monte Alban." On the theme of Monte Albán as an "information processing institution," see Marcus, "How Monte Albán Represented Itself."

15. The word in modern Hebrew is שיבולת.

16. Homer, *The Odyssey*.

17. Maria-Reina Bravo, "Doris Salcedo, Shibboleth."

18. "Shibboleth," *Monster Wiki*, https://monster.fandom.com/wiki/Shibboleth.

19. According to the shibboleth.net website: "With just one identity, a user can securely sign into a variety of systems while keeping management free from the burden of maintaining a collection of usernames and passwords."

20. Augoyard and Torgue, *Sonic Experience*, 62.

21. The OED politely defines the word *Ebonics* as "African-American English." The *Urban Dictionary* says less sympathetically: "A poor excuse for a failure to grasp the basics of english [*sic*]." The *Urban Dictionary* provides an example: "Don't be tellin' me dat I can't talk good cuz I speak ebonics."

22. Baker, *Polari*. Some minority languages started in fiction, films, or television series and have gathered a base of ardent practitioners. The putative alien language of Klingon from the *Star Trek* franchise is notable among them. Okrand, *The Klingon Dictionary*.

23. Stiny, "Kindergarten Grammars."

24. According to the OED, *to text* (someone) featured only since the 1990s as a verb in everyday usage: "transitive. Telecommunications. To send (a text message) to a person, mobile phone, etc.; to send a text message to. Also intransitive: to communicate by sending text messages." On the other hand, *writing* (to write) has served as a verb as well as a noun for much longer.

25. Poovey, *A History of the Modern Fact*.

26. Ong, *Rhetoric, Romance, and Technology*; Ong, *Orality and Literacy*.

27. Ong, *Rhetoric, Romance, and Technology*, 186.

28. Ong, 186.

29. Ong, 186.

30. Critical of this reductive accounting model, Alex Pazaitisa and colleagues describe double-entry bookkeeping as representing the way value is recorded in the depersonalized industrial economy. See Pazaitisa, De Filippib, and Kostakis, "Blockchain and Value Systems in the Sharing Economy."

31. Poovey, *A History of the Modern Fact*, 63.

32. Writing is also set in relationship to speaking, which recalls Jacques Derrida's account of the cultural and ontological priority of writing; he explains this in his book *Of Grammatology*. See Coyne, *Derrida for Architects*.

33. Huber, "How to Make GPS Art." For research into the city as networks of movement, see Hillier, *Space Is the Machine*; Turner and Penn, "Encoding Natural Movement as an Agent-Based System."

34. McLuhan, *The Gutenberg Galaxy*.

35. McLuhan, 95.

36. Knuth, "Two Thousand Years of Combinatorics."

37. DuPont, "The Printing Press and Cryptography," 101.

38. DuPont, 101.

39. DuPont, 100.

40. Priani, "Raymon Llull."

41. DuPont, "The Printing Press and Cryptography," 101.

42. Lowenthal, *The Past Is a Foreign Country*, 88.

43. DuPont, "The Printing Press and Cryptography," 95.

44. Alberti, *De componendis cifris*, 180.

45. Alberti, 180.

46. Ciphertown, "How to Use the Alberti Cipher Disk Device with Method 1."

47. Singh, *The Code Book*, 126–142.

48. Singh, 126.

49. Led by mathematician Alan Turing (1912–1954). Hodges, *Alan Turing*.

50. Williams, March, and Wassell, *The Mathematical Works of Leon Battista Alberti*, 193.

51. Williams, March, and Wassell, 193.

52. DuPont, "The Printing Press and Cryptography."

53. Vitruvius, *Vitruvius: The Ten Books on Architecture*, 13.

54. Carpo, *Architecture in the Age of Printing*.

55. Alberti, *On the Art of Building in Ten Books*, 204.

56. Alberti, 310.

57. Alberti, 310.

58. A key text on the cultural role of the carnivalesque is Bakhtin, *Rabelais and His World*.

59. Rescher, "Leibniz's *Machina Deciphratoria*," 113.

60. Rescher, 110.

61. Ross, "Leibniz and the Nuremberg Alchemical Society," 223.

62. The philosopher Rene Descartes (1596–1650) was also familiar with Gaffarel. See Kirsanov, "Leibniz in Paris." I discuss this alphabet in chapter 1.

63. Look, "Gottfried Wilhelm Leibniz."

64. Leibniz, "Monadology," 7.

65. Deleuze, *The Fold*, 228.

66. This is a reference to Plato's famous cave analogy, in which the external reality of the *ideas* is perceived only fleetingly in the flickering firelight of the cave's interior; Plato, *The Republic of Plato*.

67. See, for example, Prominski and Koutroufinis, "Folded Landscapes."

68. The passage continues: "These folds, ropes, or springs set up on the opaque cloth represent innate knowledge, but an innate knowledge which passes into action when called upon by matter. For the latter unleashes the 'vibrations or oscillations' at the lower extremity of the ropes by means of 'small openings' which do exist on the lower level." Deleuze, *The Fold*, 228.

69. Look, "Gottfried Wilhelm Leibniz."

Chapter 3

1. Sebeok, *Signs*.

2. Nöth, *Handbook of Semiotics*.

3. Coyne, *Derrida for Architects*; Coyne, *Peirce for Architects*.

4. For example, see Windsor, "An Ecological Approach to Semiotics."

5. Shannon and Weaver, *The Mathematical Theory of Communication*; Reddy, "The Conduit Metaphor."

6. I am skirting past the findings of studies into language acquisition, but the view here accords with the pragmatism of Reddy, "The Conduit Metaphor." I return to this theme in chapter 13 in a discussion of languages for communicating with supposed extraterrestrial aliens.

7. Norman, *The Design of Everyday Things*.

8. Lynch and Hack, *Site Planning*, 218–219. Also see Lynch, *The Image of the City*.

9. For a critique of physical signs and how they turn places into non-places, see Augé, *Non-places*.

10. Unger and Grassl, "Insta-Holidays and Instagrammability," 92. Also see Degen and Rose, *The New Urban Aesthetic*.

11. Gehl, *Cities for People*.

12. In the *Handbook of Semiotics*, Nöth amplifies the role of "unwritten traditional rules of social conduct . . . for example the code of decorum or the fashion code." Nöth, *Handbook of Semiotics*, 206. Certain prescribed activities and conditions are mandated and presented as regulations. According to Nöth these are "sets of rules prescribing forms of social behaviour." These are typically written down and understood well in the urban context as ordinances, codes, regulations, laws, and instructions.

13. Lynch and Hack, *Site Planning*, 226.

14. In *The Image of the City*, he affirms: "A legible city would be one whose districts or landmarks or pathways are easily identifiable and are easily grouped into an overall pattern" (3).

15. Lynch and Hack, *Site Planning*, 226.

16. Lynch and Hack, 226.

17. "Yet certain elements will be crucial to all: the main system of circulation, the basic functional and social areas, the principal centers of activity and of symbolic value, the historic points, the natural site, the major open spaces." Lynch and Hack, 226.

18. Lynch and Hack, 226.

19. Lynch and Hack, 204–205.

20. Lynch and Hack, 193.

21. Lynch and Hack, 199.

22. See https://www.airbnb.co.uk/s/experiences.

23. Lynch, *The Image of the City*, 5–6.

24. Lynch, 5–6.

25. Lynch and Hack, *Site Planning*, 226.

26. Pullan, "Agon in Urban Conflict," 222.

27. Hoffman, *Cybersecurity Bible*, 604.

28. McEwen, *Vitruvius*.

29. Singh, *The Code Book*, 257.

30. Coyne, Lee, and Parker, "Permeable Portals."

31. More, *Utopia*, 65.

32. Jacobs, *The Death and Life of Great American Cities*, 32.

33. Schneier, *Beyond Fear*, 278.

34. Schneier, 278.

35. Schneier, 278.

36. OED.

37. See, for example, Lessig, *Code and Other Laws of Cyberspace*; Diver, *Digisprudence*.

38. Kitchin, "From a Single Line of Code to an Entire City," 16.

39. Kitchin, 17.

40. Graham, "Software-Sorted Geographies," 563.

41. Ben-Joseph, *The Code of the City*.

42. The philosopher Jacques Derrida makes much of concepts that revolve around *arche* and anarchy. Derrida, *Of Grammatology*. Also see Evans, "Towards Anarchitecture."

43. May, "The Crypto Anarchist Manifesto."

44. May.

45. May.

46. May.

47. Levy, "Crypto Rebels."

48. For example, the architect Bernard Tschumi's studio exercise of designing a nightclub in a graveyard: Tschumi, *Architecture and Disjunction*.

49. Mumford, *The Culture of Cities*, 288.

50. Mumford, 288.

51. Manaugh, *A Burglar's Guide to the City*.

52. Manaugh, 11–12.

53. Manaugh, 11–12.

54. Connor, *Beyond Words*, 48. Indeed, a search on my Apple music streaming service calls up many songs and albums simply titled "Whisper." There's George Michael's "Careless Whisper" sung in full voice. A rap song called "Wait (The Whisper Song)" by Ying Yang Twins starts, "Hey, how you doin' lil' mama? Let me whisper in your ear. Tell you somethin' that you might like to hear." That's a rap song delivered as a whisper—appropriate to its transgressive lyrics.

55. Known as ASMR (autonomic sensory meridian response) videos, designed to invoke a state of deep relaxation and sometimes euphoria through a putative psychophysical "auto sensory meridian response." For further explanation of ASMR, see my blog post: Coyne, "The Pleasures of the Mouth."

56. I review some of these considerations in Coyne, *The Tuning of Place*.

57. Connor, *Beyond Words*, 50.

58. Connor, 50–51.

59. Connor, 51.

60. See Tanizaki, *In Praise of Shadows*. Whispers belong in the half light of a private room, a building's eaves (where *eavesdroppers* hang out), in colonnades, doorways, and other thresholds.

61. See Crandall, "Invisible Commercials and Hidden Persuaders."

62. Packard, *The Hidden Persuaders*, 31.

63. Brean, "'Hidden Sell' Technique Is Almost Here."

64. In fact, that particular edition of *Life* magazine contains many other "subliminal messages." In keeping with the age, the edition purveys the good life, middle-class suburbia, a burgeoning consumer culture, gender stereotypes and the nuclear family—wash and wear shirts, lawn mowers, pasta sauce, washing machines, *South Pacific*, a Smith-Corona portable electric typewriter, a Keystone 8mm movie camera.

Chapter 4

1. Mumford, *The Culture of Cities*, 3.

2. That's 10×9×8×7×6×5×4×3×2×1 orderings, or 10! (10 factorial).

3. Technically, some of these associations are "false cognates." According to the *Online Etymological Dictionary*, "hack" applied to labor derives from the pastureland in Hackney, England, where horses were kept for hire; https://www.etymonline.com/search?q=hack.

4. Shields, *Collage and Architecture*, 2.

5. Ellison, "Deciphering and the Exhaustion of Recombination," 181.

6. "A cabinet is redesigned so that visible drawers, already built to hide away documents, contain additional hidden compartments. Similarly, a book or a sentence can be restructured in ways that bend the conventions of their use, disguising messages in folds or between the lines." Ellison, 181.

7. Kurokawa, *Metabolism in Architecture*.

8. Anon., "TOOOL," https://toool.nl/Toool; Kurokawa, *Metabolism in Architecture*.

9. These are specialists who select locks from a catalogue and work out which locks share the same key, and master key.

10. Herbert Simon, theorist of decision-making, coined the term *satisficing* to account for the challenge of deciding between options in the face of multiple and conflicting criteria. See Simon, "The Structure of Ill-Structured Problems"; Newell and Simon, *Human Problem Solving*; Simon, *The Sciences of the Artificial*.

11. Conan Doyle, *The Sign of Four*, 42.

12. Descartes, *Discourse on Method and the Meditations*, 41.

13. Alexander, *Notes on the Synthesis of Form*; Jones, *Design Methods*.

14. Also see Steadman, *Architectural Morphology*.

15. Bloch and Krishnamurti, "The Counting of Rectangular Dissections."

16. Bloch and Krishnamurti.

17. Coyne, *Logic Models of Design*.

18. Designers don't design in this way. How designers might actually design, how elements get laid out in a city, and citizens organize space, is beyond the scope of this chapter. For an account of the difficulties with logical models of design and my retelling in the context of the pragmatic philosophy of Charles Sanders Peirce, see Coyne, *Peirce for Architects*.

19. Ong, *Ramus*.

20. See Lefebvre, *Rhythmanalysis*.

21. Descartes, *Discourse on Method and the Meditations*.

22. *Squid Game* (2021), Series 1, Episode 7: "VIPS."

23. A riddle falls within the purview of humor as a kind of joke. See Martin and Ford, *The Psychology of Humor*, 4.

24. That's close to a description of Salvador Dalí's painting *The Temptation of St Anthony*.

25. Topolinski and Reber, "Gaining Insight into the 'Aha' Experience."

26. Lyons, "English Letter Frequencies."

27. ETA, "Letter Frequencies."

28. Ohlman, "Subject-Word Letter Frequencies with Applications to Superimposed Coding."

29. Ohlman, "Subject-Word Letter Frequencies," 903.

30. Though it is likely a computer would produce a range of possible plain text messages from that short cryptogram. Mia Epner provides a helpful video explaining encryption keys. See Epner, "Encryption and Public Keys." For further explanation, see Lake, "What Is AES Encryption and How Does It Work?"; Watson, "Famous Codes and Ciphers through History."

31. Watson, "Famous Codes and Ciphers through History."

32. Sullivan, "A (Relatively Easy To Understand) Primer on Elliptic Curve Cryptography."

33. Mann, "The Science of Encryption."

34. I joined the cadre of bloggers who seek to explain and clarify cryptography with my own contribution; see R. Coyne, "RSA Public Key Encryption."

35. Seetharam, "RSA (Rivest, Shamir, Adleman) Algorithm Explained with Example."

36. Hughes, "How Elliptic Curve Cryptography Works." Diffie and Hellman were the computer scientists who identified the challenge of asymmetrical key encryption; see Diffie and Hellman, "New Directions in Cryptography."

37. Coyne, "Key Exchange"; Coyne, "Elliptic Trapdoors"; Coyne, "Elliptic Fields."

38. Here I follow the verdict of many linguists and theorists of metaphor that metaphors permeate every area of life and practice, including mathematics. See Black, *Models and Metaphors*; Lakoff and Johnson, *Metaphors We Live By*.

39. Aristotle introduced one of the main ideas about metaphor, that "a metaphor is the application of a noun that properly applies to something else." See Aristotle, *Poetics*, 34.

40. Manaugh, *A Burglar's Guide to the City*, 76.

41. Manaugh, 76.

42. Manaugh, 76.

43. Manaugh, 273.

Chapter 5

1. Langewiesche, "Welcome to the Dark Net."

2. Langewiesche.

3. Dingledine, Syverson, and Mathewson, "Brows Privately. Explore Freely."

4. Anderson, *Imaginary Cities*, 23.

5. Anderson, 19.

6. "In what might be termed a 'vertical turn', the politics of subterranea is a topic that a range of thinkers have turned increasing attention to." Dobraszczyk, Galviz, and Garrett, *Global Undergrounds*, 15.

7. Manaugh, *A Burglar's Guide to the City*, 274.

8. McCall Smith, *44 Scotland Street*.

9. Dobraszczyk, Galviz, and Garrett, *Global Undergrounds*, 11.

10. Anderson, *Imaginary Cities*, 21.

11. Negarestani, *Cyclonopedia*, 66.

12. Douglas, *Purity and Danger*.

13. Borges, "The Immortal," 104.

14. Eco, *Reflections on the Name of the Rose*.

15. Eco, *The Name of the Rose*.

16. Eco, *Reflections on the Name of the Rose*, 57.

17. Such markets are usually in fact well organized, with sellers of produce allotted to their own areas. Proponents of unregulated software development in the sharing economy drew on the myth of the chaotic bazaar. See Raymond, *The Cathedral and the Bazaar*.

18. Eco, *Reflections on the Name of the Rose*, 58.

19. Pennick, *Mazes and Labyrinths*.

20. Pennick, 18.

21. Caillois, *Man, Play and Games*, 23.

22. Pennick, *Mazes and Labyrinths*, 39.

23. Examples include Daniel Libeskind and Cosentino's "Musical Labyrinth" project in Frankfurt, which is marked out as white lines and text on the ground plane of a public square. Libeskind's Jewish Museum in Berlin features a grid of obelisks described on the museum's website (https://www.jmberlin.de): "The labyrinthine 'Garden of Exile' tests the visitor's sense of balance and provides a metaphor for the loss of orientation in foreign countries." It is a 7×7 grid. The base and the rectilinear obelisks are at slight angles to the ground plane.

24. Douglas, *Purity and Danger*, 44.

25. To reinforce this connection, note how Negarestani refers to the Mesopotamian giant Humbaba, whose "labyrinthine face (with unicursal human entrails as the beard) recalls the early art of Haruspicy (divination using the liver or entrails) in ancient Mesopotamian cultures." Negarestani, *Cyclonopedia*, 115.

26. Negarestani, 43.

27. Falahat, *Re-imaging the City*.

28. Mumford, *The Culture of Cities*, 291–292.

29. Described in Negarestani, *Cyclonopedia*, 51.

30. He refers to the book as a work of philo-fiction. In a helpful explanation in a blog by Terence Blake the prefix philo- refers to the compulsory course in French schooling known casually as *philo*. The genre is also known as *theory-fiction*.

31. Borges, "The Aleph," 423.

32. The panopticon plan was championed by the nineteenth-century utilitarian social reformer Jeremy Bentham (Bentham, *The Works of Jeremy Bentham*, vol. 4) and critiqued as a model of social control by Michel Foucault (Foucault, *Discipline and Punish*), and many urban critics have adopted it as leitmotif for the surveillance society (e.g., Levin, Frohne, and Weibel, *CTRL [SPACE]*).

33. Borges, "The Aleph," 423.

34. Borges, "Library of Babel," 59.

35. Borges, 58.

36. Weinsheimer, *Gadamer's Hermeneutics*.

37. The repeated near-far experience also resonates with Freud's account of the uncanny. See Freud, "The 'Uncanny.'"

38. Batty and Hudson-Smith, "Imagining the Recursive City," 39.

39. Batty and Hudson-Smith.

40. Hofstadter and Dennett, *Gödel, Escher, Bach*, 135.

41. Hofstadter and Dennett, 136.

42. Hodge, "What Is Tor? A Beginner's Guide to Using the Private Browser."

43. Hofstadter and Dennett, *Gödel, Escher, Bach*.

44. Hofstadter and Dennett, 136.

45. For a political take on the concept of "the stack," see Bratton, *The Stack*.

46. Weinsheimer, *Gadamer's Hermeneutics*, 104.

Chapter 6

1. Baudrillard, *Simulacra and Simulation*. For many critical theorists, the reality of the city is capitalist hegemony and consumerism, and the inequalities and injustices they perpetuate. Contemporary critiques of the hidden biases of digital culture, as well as racial, gender, age, and other assumptions and discriminatory practices, include Williams, *Data Action*, and Benjamin, *Race After Technology*.

2. Caillois, *Man, Play and Games*, 19.

3. Caillois, 22.

4. For a critique of some UX design, see Gray et al., "The Dark (Patterns) Side of UX Design."

5. Whitson, "Gaming the Quantified Self."

6. Mac Sithigh and Siems, "The Chinese Social Credit System."

7. Huizinga, *Homo Ludens*.

8. "Emma Watson First Day in the Office," scene from the film *The Circle* (dir. James Ponsoldt, 2018), https://www.youtube.com/watch?v=0ByQ35OLoMM.

9. Nicholson, "A Recipe for Meaningful Gamification," 4.

10. Nicholson, 4.

11. The answer is "nothing."

12. For an account of the likely impact of quantum computing on cryptography, see Martin, *Cryptography*, 321.

13. Ellison, *A Material History*.

14. The film *Jumanji: Welcome to the Jungle* (dir. Jake Kasdan, 2017) uses this map scenario.

15. Brown, *The Lost Symbol*, 234.

16. https://www.dyne.org.

17. Shamir, "How to Share a Secret."

18. Anon., *Digital Government*.

Chapter 7

1. Mumford called this primal urban condition the "eopolis." See Mumford, *The Culture of Cities*; Geddes, *Cities in Evolution*. For related explanations of "social economics" and the gift society, see my own book on the subject: Coyne, *Cornucopia Limited*.

2. See Daniels Trading, "Bitcoin: Commodity or Currency"; Zuboff, *The Age of Surveillance Capitalism*.

3. Godbout, *The World of the Gift*.

4. Rivera, "Potential Negative Effects of a Cashless Society."

5. Jacobs, *The Death and Life of Great American Cities*, 421.

6. Nakamoto, "Bitcoin," 1.

7. Canellis, "Bitcoin Has Nearly 100,000 Nodes."

8. Anon., "SHA-256 Hash Calculator."

9. I alluded to the ubiquitous *block* idea in chapter 2 in relation to city blocks and printing blocks. The idea of the *blockchain* also bears some similarities to the *stack* described in chapter 5 in reference to keeping track of recursion and sequential data structures.

10. Anon., "Bitcoin Mining Guide." Also see Sutton, *Cryptocurrency Mining*.

11. Cannon and Tuwiner, "Is Bitcoin Mining Profitable or Worth It in 2020?"

12. These figures are frequently monitored and reported in the press; see, for example, Kharpal, "A Major Chinese Bitcoin Mining Hub Is Shutting Down Its Cryptocurrency Operations."

Chapter 8

1. Taylor, "The World's First Bitcoin Republic."

2. Elnagar, "IMF Executive Board Concludes 2021 Article IV Consultation with El Salvador."

3. For notes on the value of Bitcoin, see Edwards, "Bitcoin's Price History."

4. Remitly, https://www.remitly.com/us/en.

5. De Vries, "Bitcoin's Growing Energy Problem."

6. Naderzadeh, *Cryptocurrency Mining Using Renewable Energy*, 4. See also media accounts of "volcano-powered bitcoin mining" in El Salvador: Sigalos, "El Salvador Has Just Started Mining Bitcoin Using the Energy from Volcanoes." The president of El Salvador commissioned Mexican architect Fernando Romero (https://fr-ee.org) to design Bitcoin City adjacent to the volcano.

7. This is currently being challenged by increasingly centralized trading exchanges such as Coinbase and Jaxx, and Bitcoin's new status as a pseudo-commodity; see Daniels Trading, "Bitcoin: Commodity or Currency."

8. Saleh, "Blockchain without Waste," 1162. Also see Kiayias et al., "Ouroboros."

9. Saleh, 1176; Kiayias et al., "Ouroboros." See also https://cardano.org.

10. Bezek, "What Is Proof-of-Stake, and Why Is Ethereum Adopting It?"

11. Shiller, "The Bitcoin Narratives."

12. Szabo, "Formalizing and Securing Relationships on Public Networks"; Dewan and Singh, "Use of Blockchain in Designing Smart City."

13. Sundararajan, *The Sharing Economy*, 93.

14. Sundararajan. Contracts are verified with "zero knowledge" of content.

15. Sheldon, "Auditing the Blockchain Oracle Problem."

16. Oliva, Hassan, and Jiang, "An Exploratory Study of Smart Contracts."

17. Anon., "How Blockchain Technology and Smart Contracts Are Transforming the Real Estate Industry."

18. Shojaei et al., "An Implementation of Smart Contracts by Integrating BIM and Blockchain," 519. Colleagues and I have explored the application of smart contracts to 3D models in virtual reality (VR) environments; see Coyne, "Transactions in Virtual Places." For a different perspective, see Dounas, Lombardi, and Jabi, "Framework for Decentralised Architectural Design BIM and Blockchain Integration."

19. A smart contract could respond to triggers such as a transaction has been approved or completed, a certain amount of time has elapsed or a date is reached, the goods are at the optimal price for the purchaser, an item to be purchased is of a particular (approved) type (e.g., you can spend this cryptocurrency to buy an airline ticket but not on entertainment), or the item in the transaction has been resold. Here are some possible actions following such triggers: the smart contract prevents or enables further transactions, distributes funds to other parties, awards a refund, imposes a fee or penalty, sends an email, issues a private key to unlock a building, or switches on the heating.

20. For a discussion of big data in the urban context, see Engin et al., "Data-Driven Urban Management."

21. In his *Utopia*, the Tudor scholar Thomas More said that people behave themselves as "everyone has his eye on you." The ideal city was designed without places to hide.

22. Golumbia, "Bitcoin as Politics," 119.

23. For the application of distributed ledgers in this context, see Townsend, *Distributed Ledgers*.

24. Naderzadeh, "Cryptocurrency Mining Using Renewable energy," 5.

25. Reiff, "How to Identify Cryptocurrency and ICO Scams."

26. For a further example, see Tavernise and Yaffe-Bellany, "Death of a Crypto Company."

27. Frankel, "An Anonymous Left-wing Art Group Known in the 1990s as Luther Blissett."

28. Kaminska, "From Bitcoin to QAnon."

29. Kaminska.

Chapter 9

1. Degen and Rose, *The New Urban Aesthetic*, 41.

2. E.g., Base, *Animalia*.

3. Young Rival, "Black Is Good [Official] (Autostereogram Video)."

4. Roma N, "Animated Stereogram."

5. Desmedt, Hou, and Quisquater, "Audio and Optical Cryptography."

6. Bowen, "ASCII Stereograms."

7. Naor and Shamir, "Visual Cryptography," 1.

8. To illustrate the method Naor and Shamir provided two images: "two random looking dot patterns. To decrypt the secret message, the reader should photocopy each pattern on a separate transparency, align them carefully, and project the result with an overhead projector." Naor and Shamir, "Visual Cryptography," 1.

9. Petrauskiene and Saunoriene, "Application of Dynamic Visual Cryptography."

10. Desmedt, Hou, and Quisquater, "Audio and Optical Cryptography."

11. Vitruvius, *Vitruvius: Ten Books on Architecture*.

12. Alberti, *On the Art of Building in Ten Books*.

13. Carpo, *Architecture in the Age of Printing*. For all his architectural inventiveness, some scholars describe Alberti as an iconophobe.

14. Maier, *Rome Measured and Imagined*.

15. I provide an illustration of pixel manipulation to blur and image on my blog: Coyne, "Urbanise Rasterise."

16. Rowe and Koetter, *Collage City*.

17. Anon., "AEC (UK) CAD Standard for Basic Layer Naming."

18. Tanna, *Codes, Ciphers, Steganography and Secret Messages*; Al-Mousaoui, *Image*; Bailey and Curran, *Steganography*.

19. Europol's European Cyber Crime Centre has an initiative to monitor and prevent the criminal use of information hiding techniques. See https://cuing.org.

20. Loos, *Ornament and Crime*.

21. I provide a worked example of such concealment in a blog post: Coyne, "Ornament and Crime."

22. In this sense, crypto art is no more a genre, subgenre, style, or movement in art than auction art, gallery art, or collectible art.

23. Hexeosis, "Why I Sell My Gifs as Crypto Art"; Hexiosis, "About Crypto Art." Also see Kim, "Mars House."

24. Abrahamson, "An NFT Just Sold for $69,346,250."

25. Hall and Watkinson, "CryptoPunks."

26. Marx described this substitution as "a process of analysis—by subdivisions of labour, which transforms the worker's operations more and more into mechanical operations, so that, at a certain point, the mechanism can step into his place." Marx, *Karl Marx*, 379.

27. Mitchell, *The Reconfigured Eye*, 51.

28. There is scope for a crypto art that expands from its mode of transaction in cryptocurrency to the arena of cryptography, encryption, and cryptanalysis. Moving crypto art beyond the borders of the canvas and the flat screen to performance art would extend the crypto theme further.

29. Hollander, "Why NFTs Matter to Urban Planning."

Chapter 10

1. Bower and Green, "STIS Records a Black Hole's Signature."

2. Robinson, Boyd, and Fetterman, "An Emotional Signature of Political Ideology," 98. Such observations indicate that signatures can be as expansive and difficult to identify as emotional states. Here it's really a particular word that provides the signature: "fear" summarizes a conservative agenda, and "anger" accounts for liberals.

3. For a discussion of emotion in urban and digital contexts, see Coyne, "Melancholy Urbanism."

4. Furno et al., "A Tale of Ten Cities."

5. Degen and Rose, *The New Urban Aesthetic*, 77. Detecting flows is a bit like finding the rapid flow of material around a black hole to determine the black hole's presence and size, even though you can't see it.

6. To reduce an image to 1/16 its size on the screen, a reduction algorithm divides the image into 4×4 pixel squares and displays the average of the color value of each pixel in the 4×4 square as a single pixel in a new file to create the new reduced image.

7. The most common method uses an algorithm that creates a color table or palette of the colors that appear in an image.

8. One of the simplest ways to produce a hash of an image is to reduce its color range to a grayscale, then reduce its size repeatedly by the averaging process described previously to something as small as 8×8 pixels. See Anon., "Testing Different Image Hash Functions."

9. The ubiquitous hash symbol, the octothorp (#), has two pairs of lines that cut across each other, and as a hashtag has come to indicate a means of cross-referring to another part of a document.

10. Here I am using an online tool at https://www.browserling.com/tools/all-hashes.

11. Lawrence, "Graffiti Strategy for Edinburgh."

12. Ferguson and Schneier, *Practical Cryptography*, 83–95; Madeira, "How Does a Hashing Algorithm Work?"

13. Hall and Watkinson, "CryptoPunks."

14. I used a drag and drop tool for this at GitHub: Anon., "SHA256 File Checksum."

15. Douglass, "An Introduction to IPFS."

16. Merkle, "A Digital Signature Based on a Conventional Encryption Function."

17. Lewis, "A Gentle Introduction to Blockchain Technology"; Lewis, "A Gentle Introduction to Immutability of Blockchains"; Lewis, *The Basics of Bitcoins and Blockchains*.

18. Note the creation of the "permaweb" that deploys blockchain technology to render certain web content permanent. See https://www.arweave.org.

19. All Hash Generator, https://www.browserling.com/tools/all-hashes.

20. Lewis, "A Gentle Introduction to Immutability of Blockchains." There are several other sophistications to the puzzle involving a range of terms such as *cryptographic nonce*. The nonce is the number appended to an encrypted block. Miners need to work out what that number is in order for the block to generate a hash that meets the criterion of the leading 0s.

21. Bispo, "Cryptocurrency, Graphics Card Shortage, and the Rise of Cloud GPU."

22. Nakamoto, "Bitcoin," 3.

23. Kiayias et al., "Ouroboros."

Chapter 11

1. That's my own example. I draw in this discussion on Brunton and Nissenbaum, *Obfuscation*. They give the example of the way spiders confuse predators.

2. What Friedrich Nietzsche describes as "the formless unformulable world of the chaos of sensations." Nietzsche, *The Will to Power*, 307.

3. Brunton and Nissenbaum, *Obfuscation*. Also see Gray et al., "The Dark (Patterns) Side of UX Design."

4. Brunton and Nissenbaum, *Obfuscation*, 7.

5. Brunton and Nissenbaum, 30.

6. Also see Gray et al., "The Dark (Patterns) Side of UX Design."

7. Mueller, *Report on the Investigation into Russian Interference in the 2016 Presidential Election*.

8. Manchester, "Mueller Report Suggests Trump Intended to Obstruct Investigation, Says Ex-Watergate Prosecutor."

9. For a critique of Google's index and search, see Vaidhyanathan, *The Googlization of Everything*.

10. See an account in the book by the chair of the United States House Permanent Select Committee on Intelligence: Schiff, *Midnight in Washington*.

11. Haines, *Foreign Threats to the 2020 US Federal Elections*.

12. DiResta et al., *The Tactics and Tropes of the Internet Research Agency*; Howard et al., *The IRA, Social Media and Political Polarization in the United States, 2012–2018*.

13. DiResta et al., *The Tactics and Tropes of the Internet Research Agency*, 8. Also see Schiff, *Midnight in Washington*.

14. Surowiec, "Post-truth Soft Power." For compelling accounts of Russian influence in U.S. politics, see Nance, *The Plot to Betray America*, and Nance, *They Want to Kill Americans*.

15. Surowiec, 25.

16. Surowiec, 24.

17. Nicole Perlroth outlines the parlous state of cybersecurity defenses in cities and states. See Perlroth, *This Is How They Tell Me the World Ends*.

18. Wicker, "The Ethics of Zero-Day Exploits," 99.

19. Wicker, 99.

20. Perlroth, *This Is How They Tell Me the World Ends*.

21. Suiche, "Shadow Brokers."

22. Snowden, *Permanent Record*.

23. For Snowden's own account, see *Permanent Record*. This was followed up by the 2016 film *Snowden*, directed by Oliver Stone.

24. Snowden, *Permanent Record*.

25. Greenwald, "Edward Snowden"; Greenwald, "NSA Collecting Phone Records of Millions of Verizon Customers Daily."

26. Bernal, "Does the UK Engage in 'Mass Surveillance'?"

27. Alerts to the issue of bulk data collection are also provided in Moody, "What's the Difference Between 'Mass Surveillance' and 'Bulk Collection'?"; MI5, "Introduction to Bulk Data."

28. Greenberg, "How Not to Prevent a Cyberwar with Russia."

29. Perlroth and Shane, "In Baltimore and Beyond, a Stolen N.S.A. Tool Wreaks Havoc." Also see Perlroth, *This Is How They Tell Me the World Ends*.

Chapter 12

1. See Popov et al., "Apply Magic Sauce." I fed this paragraph through the system and it inferred that I as the author am "liberal and artistic," contemplative, competitive, relaxed, etc.

2. Dang and Goller, "What They Said."

3. Dang and Goller.

4. Zuboff, *The Age of Surveillance Capitalism*.

5. Benjamin, *Race After Technology*, 16. For a reading of the relationship between ethics and digital systems, see Amoore, *Cloud Ethics*. I think the political and ethical challenges posed by machine learning algorithms extends to every kind of algorithm.

6. Sommerville, *Software Engineering*.

7. Benjamin, *Race After Technology*, 16.

8. The area of bias is addressed by studies into interpretation and hermeneutics. See, for example, Gadamer, *Truth and Method*; Snodgrass and Coyne, *Interpretation in Architecture*.

9. Accessibility is written into law in many countries. See, for example, the UK Equality Act 2010, https://www.legislation.gov.uk/ukpga/2010/15/section/20.

10. CDRC Mapmaker, https://maps.cdrc.ac.uk/.

11. Deas et al., "Measuring Neighbourhood Deprivation."

12. Amoore, *Cloud Ethics*, 6.

13. Amoore, 109.

14. Amoore, 8.

15. Howe, Zer-Aviv, and Nissenbaum, "AdNauseam."

16. Howe, Toubiana, and Nissenbaum, "TrackMeNot."

17. Brunton and Nissenbaum, *Obfuscation*. A low-tech method is to share SIM cards and loyalty cards to confound profilers. This is a bit like masking confidential in-person conversations by meeting in a noisy bar, or a place where there are many voices to drown out your own. In this section I have put the spotlight on algorithms, the design and implementation of which inevitably interact with developments in

legislation such as the highly consequential GDPR in Europe and similar in other parts of the world. See https://gdpr.eu.

18. See Rumelhart and McClelland, *Parallel Distributed Processing*. Using such methods, natural language processing (NLP) programs now seem able to take as input words and phrases provided by a human operator and generate coherent sentences in response, simulating a kind of dialogue that scholars such as David Chalmers think exhibit general intelligence. See Weinberg, "Philosophers on GPT-3 (updated with replies by GPT-3)." For my own experimentation with the GPT-3 based AI platform, see Coyne, "Benefits of Artificial Misinformation."

19. Vision AI, https://cloud.google.com/vision. Also see https://openai.com/dall-e-2/.

20. The latter includes Google's reCAPTCHA tool that supports login routines that ask you to prove you are a human user and not a malicious program by giving you a simple pattern-matching test. The data from that test feeds the training algorithm. See https://www.google.com/recaptcha/about/.

21. For a review of feature detection in the case of landscape assessment, see Wilkins et al., "Promises and Pitfalls of Using Computer Vision."

22. Amoore, *Cloud Ethics*, 71.

23. Amoore, 172.

24. Das et al., "Estimating Likelihood of Future Crashes for Crash-Prone Drivers."

25. Loo et al., "Applying the Hidden Markov Model"; Rabiner and Juang, "An Introduction to Hidden Markov Models."

26. Jeon, Jordan, and Krishnamoorthy, "On Modeling ASR Word Confidence." I also explain the approach at Coyne "Speech to Text." For a discussion of race differences and automated speech recognition, see Benjamin, *Race After Technology*, 15.

27. Ahmed, Natarajan, and Rao, "Discrete Cosine Transform," 93. I provide a worked example at Coyne, "COVID-19 Rhythmanalysis."

28. Lefebvre, *Rhythmanalysis*.

29. Coyne, *The Tuning of Place*.

30. Nevejan, Sefkatli, and Cunningham, *City Rhythm*.

31. Nevejan, Sefkatli, and Cunningham, 3.

32. Nevejan, Sefkatli, and Cunningham, 3.

33. Nevejan, Sefkatli, and Cunningham, 4.

34. Nevejan, Sefkatli, and Cunningham, 44.

35. The authors provide a diagram that shows "the rhythms of a fictional elderly persona (yellow), rhythms of a young fictional persona (red) and rhythms of the shopping centre (green)." Nevejan, Sefkatli, and Cunningham, 47.

36. Nevejan, Sefkatli, and Cunningham, 74.

37. Neuman, "Keeping Secrets."

38. Karpel, "Family Secrets," 2.

39. Manaugh, *A Burglar's Guide to the City*, 263.

40. The website faketv.com bears the heading "Burglars Love Easy Targets."

Chapter 13

1. Bosworth, "Building the Metaverse Responsibly." The term *metaverse* was coined by Stephenson in the 1992 novel *Snow Crash*. For a discussion of the relationship between *metaverse* and *multiverse*, see Dallow, "The Media Multiverse and Adaptive Virtuality."

2. Ito, Okabe, and, Matsuda, *Personal, Portable, Pedestrian.*

3. Hall, *The Hidden Dimension*, 53.

4. Hall, 53.

5. LaBelle, *Background Noise.*

6. LaBelle, 325.

7. Descartes, *Discourse on Method and the Meditations*. Axes xyz offer three dimensions.

8. Or interchange the axes to rotate the object so described.

9. See, for example, Anon., "The Hardest Trip." Also see Gleick, *Chaos.*

10. I am quoting Wittgenstein again. See chapter 3.

11. Calvino, *Invisible Cities.*

12. Anderson, *Imaginary Cities.*

13. On the idea of a virtual reality metaverse, see Wright et al., "Augmented Duality."

14. The sci-fi film *Valerian and the City of a Thousand Planets* (dir. Luc Besson, 2017) offers a compelling live action and CGI rendering of a multilayered city superimposed over a desert experienced by the occupants via head-mounted displays. The film *Ready Player One* (dir. Steven Spielberg, 2018) offers a similar scenario. For a

philosophical treatment of multiple worlds, virtual reality, and the simulation hypothesis, see Chalmers, *Reality+*.

15. Abbott and Ehlinger, "Flatland—the Film"; Abbott, *Flatland*.

16. A (magic) goose explains other worlds: "millions of other universes exist unaware of one another. . . . We are as close as a heartbeat, but we can never touch or see or hear these other worlds except in the Northern Lights." Pullman, *Northern Lights*, 187.

17. Kessler, "Trump's Day on Twitter."

18. Vilenkin and Tegmark, "The Case for Parallel Universes."

19. String theory is explained in many popular science publications, including Hawking, *A Brief History of Time*.

20. These are themes developed in theology, psychology, and anthropology. See Eliade, *The Two and the One*; Turner, *The Forest of Symbols*.

21. According to sound theorist Douglas Kahn. See Kittler, von Mücke, and Similon, "Gramophone, Film, Typewriter," 12; Kahn, *Noise, Water, Meat*.

22. This philosophy of disembodiment was embraced by Plotinus (203–270 AD). Plotinus, *The Essence of Plotinus*, 55.

23. Kittler, von Mücke, and Similon, "Gramophone, Film, Typewriter," 12.

24. Enns, "Spiritualist Writing Machines," 4.

25. Conan Doyle, *The History of Spiritualism*, vol. 1, 62–62.

26. Enns, "Spiritualist Writing Machines," 4.

27. See Relph, "Spirit of Place and Sense of Place in Virtual Realities."

28. Here I draw on explanations provided by Cui et al., "An Encryption Scheme Using DNA Technology"; Dey et al., "DNA Origami"; Zahid et al., "DNA Nanotechnology."

29. Dey et al., "DNA Origami," 12.

30. Dey et al., 14.

31. Dey et al., 15.

32. Anon., "Fact Check."

33. Zhang et al., "DNA Origami."

34. Zhang et al., 2.

35. Zhang et al., 7.

36. Kahn, *The Codebreakers*, chap. 26, Kindle. For insight into interplanetary communication, I draw on this chapter, which is titled "Messages from Outer Space." Also see Oberhaus, *Extraterrestrial Languages*.

37. Freudenthal, *Lincos*.

38. Freudenthal, 145.

39. Kahn, *The Codebreakers*, chap. 26, Kindle.

40. The society website is at www.bis-space.com/.

41. Galton, "Intelligible Signals between Neighbouring Stars."

42. Galton, 658.

43. Kahn, *The Codebreakers*, chap. 26, Kindle.

44. Kahn, *The Codebreakers*, chap. 26, Kindle. SETI, the Search for Terrestrial Life Institute, continues the search for extraterrestrials by other means; see https://www.seti.org.

45. Anon., "How to Build a Quantum Internet"; Castelvecchi, "The Quantum Internet Has Arrived (and It Hasn't)."

46. For a history and examination of the two-slit experiment, see Barad, *Meeting the Universe Halfway*.

47. Gregory, *Inventing Reality*; Barad, *Meeting the Universe Halfway*.

48. For a helpful account, see Smith, "Quantum Instruction Set—Computerphile."

49. Anon., "How to Build a Quantum Internet."

50. Crane, "China's Quantum Satellite Helps Send Secure Messages over 1200km."

51. I draw on some of Robert Smith's explanations of the mathematics here. See Smith, "Quantum Instruction Set—Computerphile."

52. Crane, "China's Quantum Satellite Helps Send Secure Messages over 1200km."

53. Also see Bohm and Peat, *Science, Order and Creativity*; Gell-Mann, *The Quark and the Jaguar*; Schade, *Free Will and Consciousness in the Multiverse*.

54. Everett, "'Relative State' Formulation of Quantum Mechanics," 459. Also see Rovelli, "Relational Quantum Mechanics."

55. Everett, "'Relative State' Formulation of Quantum Mechanics," 460.

56. Everett, 462.

57. Barad, *Meeting the Universe Halfway*.

58. Interactive films such as *Bandersnatch* (dir. David Slade, 2018) in the TV series *Black Mirror* present alternate storylines, as do many video games.

59. Gregory, *Inventing Reality*.

60. Bohr, "The Bohr-Einstein Dialogue," 139.

61. Heisenberg, *Physics and Philosophy*, 46.

62. Barad, *Meeting the Universe Halfway*, 335.

63. Barad, 335.

64. Barad, 154.

65. Barad, 138.

Bibliography

Abbott, Edwin A. *Flatland: A Romance of Many Dimensions*. Boston: Roberts Brothers, 1885.

Abbott, Edwin A., and Ladd Ehlinger. "Flatland—the Film." YouTube, April 3, 2012. Accessed November 14, 2020. https://www.youtube.com/watch?v=Mfglluny8Z0.

Abrahamson, Kale. "An NFT Just Sold for $69,346,250." *The Kale Letter*, March 11, 2021. Accessed March 14, 2021. https://thekaleletter.substack.com/p/an-nft-just-sold -for-69346250?r=78pcb&utm_campaign=post&utm_medium=web&utm_source =copy.

Åhlén, Johan. "Pigpen Cipher (Decoder, Translator, History)." *Boxentriq*. Accessed February 27, 2021. https://www.boxentriq.com/code-breaking/pigpen-cipher.

Ahmed, Nasir, T. Natarajan, and K. R. Rao. "Discrete Cosine Transform." *IEEE Transactions on Computers* C-23 (1974): 90–93. https://doi.org/10.1109/T-C.1974.223784.

Alberti, Leon Battista. *De componendis cifris*. In *The Mathematical Works of Leon Battista Alberti*. Edited by Kim Williams, Lionel March, and Stephen R. Wassell, 169–187. Basel: Springer Basel, 2010.

Alberti, Leon Battista. *On the Art of Building in Ten Books*. Translated by Joseph Rykwert, Neil Leach, and Robert Tavernor. Cambridge, MA: MIT Press, 1996. First published in Latin in ca. 1450.

Alexander, Christopher. *Notes on the Synthesis of Form*. Cambridge, MA: Harvard University Press, 1964.

Alibasic, Armin, Reem Al Junaibi, Zeyar Aung, Wei Lee Woon, and Mohammad Atif Omar. "Cybersecurity for Smart Cities: A Brief Review." In *International Workshop on Data Analytics for Renewable Energy Integration DARE 2016: Data Analytics for Renewable Energy Integration*. Edited by Wei Lee Woon, Zeyar Aung, Oliver Kramer, and Stuart Madnick, 22–30. Switzerland AG: Springer, 2017.

Al-Mousaoui, Mohamed. *Image Steganography: The Art of Information Hiding in Images*. Amazon Digital Services, 2007. Kindle.

Amoore, Louise. *Cloud Ethics: Algorithms and the Attributes of Ourselves and Others.* Durham, NC: Duke University Press, 2020.

Anderson, Darran. *Imaginary Cities.* London: Influx Press, 2015.

Anon. "AEC (UK) CAD Standard for Basic Layer Naming." *The MicroStation Community TMC (UK).* Accessed March 21, 2021. https://aecuk.files.wordpress.com/2009/05 /aecukbasiclayernaminghandbook-v2-4.pdf.

Anon. "Affordances." *Interaction Design Foundation.* Accessed October 16, 2021. https://www.interaction-design.org/literature/topics/affordances.

Anon. "Bitcoin Mining Guide: Getting Started with Bitcoin Mining." *Bitcoinmining. com.* Accessed April 13, 2020. https://www.bitcoinmining.com/getting-started/.

Anon. "Busy Brotherly World of Freemasonry: The Ancient Fraternity Is Thriving in America." *Life* 41, no. 15 (1956): 104–122.

Anon. *Digital Government—Maintaining the UK Government as a World Leader in Serving Its Citizens Online.* UK Government Policy Paper: Department for Digital, Culture, Media, & Sport, March 1, 2017. Accessed March 22, 2021. https://www.gov.uk /government/publications/uk-digital-strategy/6-digital-government-maintaining-the -uk-government-as-a-world-leader-in-serving-its-citizens-online.

Anon. "Fact Check: Lipid Nanoparticles in a COVID-19 Vaccine Are There to Transport RNA Molecules." *Reuters,* December 5, 2020. Accessed December 21, 2021. https:// www.reuters.com/article/uk-factcheck-vaccine-nanoparticles-idUSKBN28F0I9.

Anon. "The Hardest Trip—Mandelbrot Fractal Zoom." *Maths Town.* Accessed March 13, 2021. https://www.youtube.com/watch?v=LhOSM6uCWxk.

Anon. "How Blockchain Technology and Smart Contracts Are Transforming the Real Estate Industry." *Chainlink,* June 8, 2021. Accessed July 25, 2022 https://blog .chain.link/blockchain-technology-real-estate/.

Anon. "How to Build a Quantum Internet." *Nature Video,* February 26, 2021. Accessed March 28, 2021. https://www.youtube.com/watch?v=soywlog1Fdk.

Anon. "A New Way to Experience Sir John Soane's Museum." *Sir John Soane's Museum London.* Accessed December 7, 2021. http://explore.soane.org.

Anon. "SHA256 File Checksum." *GitHub.* Accessed April 23, 2021. https://emn178 .github.io/online-tools/sha256_checksum.html.

Anon. "SHA-256 Hash Calculator." *Xorbin.* Accessed April 16, 2020. https://xorbin .com/tools/sha256-hash-calculator.

Anon. "Testing Different Image Hash Functions." *The Content Blockchain Project.* Accessed July 19, 2020. https://content-blockchain.org/research/testing-different -image-hash-functions/.

Anon. "TOOOL: The Open Organisation Of Lockpickers." Accessed May 17, 2020. https://toool.nl/Toool.

Aristotle. *The Ethics of Aristotle: The Nicomachean Ethics*. Translated by J. A. K. Thomson. London: Penguin, 1976. Written ca. 334–323 BC.

Aristotle. *Poetics*. Translated by Malcolm Heath. London: Penguin Classics, 1996.

Augé, Marc. *Non-places: Introduction to an Anthropology of Supermodernity*. Translated by John Howe. London: Verso, 1995.

Augoyard, Jean-François, and Henry Torgue. *Sonic Experience: A Guide to Everyday Sounds*. Translated by Andra McCartney and David Paquette. Montreal: McGill-Queen's University Press, 2005.

Aurigi, Alex. "No Need to Fix: Strategic Inclusivity in Developing and Managing the Smart City." In *Digital Futures and the City of Today: New Technologies and Physical Spaces*. Edited by Glenda Amayo Caldwell, Smith Carl H., and Edward M. Clift, 9–27. Bristol: Intellect, 2016.

Bailey, Karen, and Kevin Curran. *Steganography: The Art of Hiding Information*. New York: BookSurge Publishing, 2004.

Baker, Paul. *Polari: The Lost Language of Gay Men*. London: Routledge, 2002.

Bakhtin, Mikhail. *Rabelais and His World*. Translated by Hélène Iswolsky. Bloomington: Indiana University Press, 1984.

Barad, Karen. *Meeting the Universe Halfway*. Durham: Duke University Press, 2007.

Baraniuk, Chris. "The Secret Codes You're Not Meant to Know." *BBC Future*, December 17, 2015. Accessed March 9, 2021. https://www.bbc.com/future/article/20151217 -the-secret-codes-youre-not-meant-to-know.

Base, Graeme. *Animalia*. Friday Harbor, WA: Turtleback Books, 1999.

Batty, Michael, and Andrew Hudson-Smith. "Imagining the Recursive City: Explorations in Urban Simulacra." In *Societies and Cities in the Age of Instant Access*. Edited by Harvey Miller. J. Dordrecht, The Netherlands: Springer, 2007.

Baudrillard, Jean. *Simulacra and Simulation*. Translated by Sheila Faria Glaser. Ann Arbor: University of Michigan Press, 1994.

Benjamin, Ruha. *Race After Technology: Abolitionist Tools for the New Jim Code*. Cambridge: Polity Press, 2019.

Ben-Joseph, Eran. *The Code of the City: Standards and the Hidden Language of Place Making*. Cambridge, MA: MIT Press, 2005.

Bentham, Jeremy. *The Works of Jeremy Bentham*, vol. 4. London: John Bowring, 1787.

Berger, Lars T., and Krzysztof Iniewski. *Smart Grid Applications, Communications, and Security.* Hoboken, NJ: Wiley, 2012.

Bernal, Paul. "Does the UK Engage in 'Mass Surveillance'?" *Paul Bernal's Blog: Privacy, Human Rights, Law, The Internet, Politics and more,* January 15, 2016. Accessed November 27, 2019. https://paulbernal.wordpress.com/2016/01/15/does-the-uk-engage-in -mass-surveillance/.

Bezek, Ian. "What Is Proof-of-Stake, and Why Is Ethereum Adopting It? As Environmental Concerns Mount, Ethereum Is Switching to a More Energy-Efficient Protocol." *USNews,* July 14, 2021. Accessed November 6, 2021. https://money.usnews .com/investing/cryptocurrency/articles/what-is-proof-of-stake-and-why-is-ethereum -adopting-it.

Bispo, Nuno. "Cryptocurrency, Graphics Card Shortage, and the Rise of Cloud GPU." *Geek Culture,* December 26, 2021. Accessed July 22 2022. https://medium.com /geekculture/cryptocurrency-graphics-card-shortage-and-the-rise-of-cloud-gpu-1d 5501351ed3.

Black, Max. *Models and Metaphors: Studies in Language and Philosophy.* Ithaca: Cornell University Press, 1962.

Bloch, C. J., and Ramesh Krishnamurthi. "The Counting of Rectangular Dissections." *Environment and Planning B: Planning and Design* 5 (1978): 207–214.

Bogdan, Henrik, and Jan A. M. Snoek, eds. *Handbook of Freemasonry.* Leiden, Netherlands: Brill, 2014.

Bohm, David, and F. David Peat. *Science, Order and Creativity.* London: Routledge, 1989.

Bohr, Niels. "The Bohr-Einstein Dialogue." In *Niels Bohr: A Centenary Volume.* Edited by A. French, and P. Kennedy, 121–140. Cambridge, MA: Harvard University Press, 1985.

Borges, Jorge Luis. "The Aleph." In *Collected Ficciones of Jorge Luis Borges.* Edited by Andrew Hurley, 410–441. London: Penguin, 1944.

Borges, Jorge Luis. "The Immortal." In *Labyrinths: Selected Stories & Other Writings.* Edited by Donald A. Yates and James E. Irby, 102–114. New York: New Directions, 1962.

Borges, Jorge Luis. "Library of Babel." In *Labyrinths: Selected Stories & Other Writings.* Edited by Donald A. Yates and James E. Irby, 58–64. New York: New Directions, 1962.

Bosworth, Andrew. "Building the Metaverse Responsibly." *Facebook: Newsroom,* September 27, 2021. Accessed October 26, 2021. https://about.fb.com/news/2021/09 /building-the-metaverse-responsibly/.

Bowen, Jonathan. "ASCII Stereograms." *Museophile*. Accessed July 15, 2021. https://web.archive.org/web/20080517013244/http://archive.museophile.org/3d/ascii-3d.html.

Bower, Gary, and Richard Green. "STIS Records a Black Hole's Signature." *Hubblesite*, May 12, 1997. Accessed June 9, 2019. https://hubblesite.org/contents/media/images/1997/12/477-Image.html?news=true.

Bratton, Benjamin H. *The Stack: On Software and Sovereignty*. Cambridge, MA: MIT Press, 2015.

Brean, Herbert. "'Hidden Sell' Technique Is Almost Here: New Subliminal Gimmicks Offer Blood, Skulls and Popcorn to Movie Fans." *Life* 44, no. 13 (1958): 102–114.

Brown, Dan. *The Lost Symbol*. London: Corgie Books, 2010.

Brunton, Finn, and Helen Nissenbaum. *Obfuscation: A User's Guide for Privacy and Protest*. Cambridge, MA: MIT Press, 2015.

Burgess, Matt. "What Is the Internet of Things? WIRED Explains." *Wired Magazine*, February 16, 2018. Accessed October 22, 2021. https://www.wired.co.uk/article/internet-of-things-what-is-explained-iot.

Caillois, Roger. *Man, Play and Games*. New York: The Free Press of Glencoe, 1961.

Calvino, Italo. *Invisible Cities*. New York: Harcourt Brace Jovanovich, 1978.

Canellis, David. "Bitcoin Has Nearly 100,000 Nodes, But Over 50% Run Vulnerable Code." *Hard Fork: Blockchain, Cryptocurrencies, and Insider Stories by TNW*, May 6, 2019. Accessed July 12, 2022. https://thenextweb.com/news/bitcoin-100000-nodes-vulnerable-cryptocurrency.

Cannon, Malcolm, and Jordan Tuwiner. "Is Bitcoin Mining Profitable or Worth It in 2020?" Buy Bitcoin Worldwide. Last updated March 10, 2022. https://www.buybitcoinworldwide.com/mining/profitability/.

Carpo, Mario. *Architecture in the Age of Printing: Orality, Writing, Typography, and Printed Images in the History of Architectural Theory*. Cambridge, MA: MIT Press, 2001.

Castelvecchi, Davide. "The Quantum Internet Has Arrived (and It Hasn't)." *Nature* 554 (2018): 289–292.

Cecco, Leyland. "Google Affiliate Sidewalk Labs Abruptly Abandons Toronto Smart City Project." *The Guardian*, May 7, 2020. Accessed May 8, 2020. https://www.theguardian.com/technology/2020/may/07/google-sidewalk-labs-toronto-smart-city-abandoned.

Cerrudo, Cesar. "An Emerging US (and World) Threat: Cities Wide Open to Cyber Attacks (White Paper)." *IOActive Labs* (2015). Accessed October 13, 2021. https://ioactive.com/pdfs/IOActive_HackingCitiesPaper_CesarCerrudo.pdf.

Chalmers, David. *Reality+: Virtual Worlds and the Problems of Philosophy*. London: Penguin, 2020. Kindle.

Ciphertown. "How to Use the Alberti Cipher Disk Device with Method 1." YouTube, November 1, 2015. Accessed January 16, 2019. https://www.youtube.com /watch?v=4mNRU7h9Q_o.

Conan Doyle, Arthur. *The History of Spiritualism*, vol. 1. London: George H. Doran Company, 1926.

Conan Doyle, Arthur. *The Sign of Four*. London: Penguin, 2001. First published in 1889.

Connor, Steven. *Beyond Words: Sobs, Hums, Stutters and Other Vocalizations*. London: Reaktion, 2014.

Coyne, Richard. "Benefits of Artificial Misinformation." *Reflections on Technology, Media and Culture*, June 18, 2022. Accessed July 15, 2022. https://richardcoyne.com /2022/06/18/benefits-of-artificial-misinformation/.

Coyne, Richard. *Cornucopia Limited: Design and Dissent on the Internet*. Cambridge, MA: MIT Press, 2005.

Coyne, Richard. "COVID-19 Rhythmanalysis." *Reflections on Technology, Media and Culture*, August 15, 2020. Accessed November 10, 2021. https://richardcoyne.com /2020/08/15/rhythmanalysis/.

Coyne, Richard. *Derrida for Architects*. Abingdon: Routledge, 2011.

Coyne, Richard. "Elliptic Fields." *Reflections on Technology, Media and Culture*, June 12, 2021. Accessed December 1, 2021. https://richardcoyne.com/2021/06/12/more -on-elliptic-encryption/.

Coyne, Richard. "Elliptic Trapdoors." *Reflections on Technology, Media and Culture*, May 29, 2021. Accessed December 1, 2021. https://richardcoyne.com/2021/05/29 /elliptic-curve-cryptography/.

Coyne, Richard. "Key Exchange: It's a Wrap!" *Reflections on Technology, Media and Culture*, June 19, 2021. Accessed December 1, 2021. https://richardcoyne.com/2021 /06/19/its-a-wrap/.

Coyne, Richard. *Logic Models of Design*. London: Pitman, 1988.

Coyne, Richard. "Melancholy Urbanism: Distant Horizons and the Presentation of Place." In *Cinematic Urban Geographies*. Edited by François Penz and Richard Koeck, 175–188. London: Palgrave Macmillan, 2017.

Coyne, Richard. "Ornament and Crime." *Reflections on Technology, Media and Culture*, July 11, 2020. Accessed November 7, 2021. https://richardcoyne.com/2020/07/11 /ornament-and-crime/.

Coyne, Richard. *Peirce for Architects*. London: Routledge, 2019.

Coyne, Richard. "The Pleasures of the Mouth." *Reflections on Technology, Media & Culture*, December 21, 2019. Accessed March 25, 2021. https://richardcoyne.com /2019/12/21/the-pleasures-of-the-mouth/.

Coyne, Richard. "RSA Public Key Encryption." *Reflections on Technology, Media and Culture*, May 22, 2021. Accessed December 1, 2021. https://richardcoyne.com/2021 /05/22/rsa-encryption/.

Coyne, Richard. "Speech to Text." *Reflections on Technology, Media and Culture*, October 17, 2020. Accessed November 17, 2021. https://richardcoyne.com/2020/10/17 /speech-to-text/.

Coyne, Richard. "Transactions in Virtual Places: Sharing and Excess in Blockchain Worlds." In *The Phenomenology of Real and Virtual Places*. Edited by Erik Malcolm Champion, 76–93. London: Routledge, 2018.

Coyne, Richard. *The Tuning of Place: Sociable Spaces and Pervasive Digital Media*. Cambridge, MA: MIT Press, 2010.

Coyne, Richard. "Urbanise Rasterise." *Reflections on Technology, Media and Culture*, June 20, 2020. Accessed November 7, 2021. https://richardcoyne.com/2020/06/20 /urbanise-rasterise/.

Coyne, Richard, John Lee, and Martin Parker. "Permeable Portals: Designing Congenial Web Sites for the E-Society." *IADIS International Conference: e-Society* 1 (2004): 379–386.

Crandall, Kelly B. "Invisible Commercials and Hidden Persuaders: James M. Vicary and the Subliminal Advertising Controversy of 1957." HIS 4970: Undergraduate Honors Thesis University of Florida Department of History, April 12, 2006. Accessed July 13, 2020. http://plaza.ufl.edu/cyllek/docs/KCrandall_Thesis2006.pdf.

Crane, Leah. "China's Quantum Satellite Helps Send Secure Messages over 1200km." *New Scientist*, June 15, 2020. Accessed March 29, 2021. https://www.newscientist .com/article/2245885-chinas-quantum-satellite-helps-send-secure-messages-over -1200km/.

Cui, Guangzhao, Limin Qin, Yanfeng Wang, and Xuncai Zhang. "An Encryption Scheme Using DNA Technology." *IEEE 2008 3rd International Conference on Bio-Inspired Computing: Theories and Applications*. Accessed December 20, 2021. https:// ieeexplore.ieee.org/abstract/document/4656701?casa_token=2l5XHZsJmtAAAAAA: -TSm03BZXRXEV3i1zlG1V2zInAT3MP-KkCSev5Z8lnLIWsXxsZkgfkKcethTe8tKgb _fjHi8.

Curl, James Stevens. *The Art and Architecture of Freemasonry: An Introductory Study*. London: Batsford, 1991.

Curl, James Stevens. "Freemasonry and Architecture." In *Handbook of Freemasonry*. Edited by Henrik Bogdan and Jan A. M. Snoek, 557–605. Leiden, Netherlands: Brill, 2014.

Dallow, Peter. "The Media Multiverse and Adaptive Virtuality." In *Jahrbuch immersiver Medien*. Edited by Matthias Bauer, Knut Hartmann, Fabienne Liptay, Susanne Marschall, Jörg R. J. Schirra, Jörg Schweinitz, Pradeep Sen, and Hans Jürgen Wulff, 63–75. Marburg, Germany: Institut für immersive Medien (ifim) an der Fachhochschule Kiel, 2011.

Dang, Sheila, and Howard Goller. "What They Said: Quotes from a Facebook Hearing in Congress." *Reuters: Technology*, October 5, 2021. Accessed July 12, 2022. https://www.reuters.com/technology/facebook-damage-will-haunt-generation-us-senator-says-2021-10-05/.

Daniels Trading. "Bitcoin: Commodity or Currency." *StoneX*, December 12, 2017. Accessed January 11, 2018. https://www.danielstrading.com/futures-trading-education/2017/12/12/bitcoin-commodity-currency.

Das, Subasish, Xiaoduan Sun, Fan Wang, and Charles Leboeuf. "Estimating Likelihood of Future Crashes for Crash-Prone Drivers." *Journal of Traffic and Transportation Engineering* 2, no. 3 (2015): 145–157.

Deakin, Mark, and Alasdair Reid. "Smart Cities: Under-gridding the Sustainability of City-Districts as Energy Efficient-Low Carbon Zones." *Journal of Cleaner Production* 173 (2018): 39–48. http://dx.doi.org/10.1016/j.jclepro.2016.12.054.

Deas, Iain, Brian Robson, Cecilia Wong, and Michael Bradford. "Measuring Neighbourhood Deprivation: A Critique of the Index of Multiple Deprivation." *Environment and Planning C: Government and Policy* 21, no. 6 (2003): 883–903. https://doi.org/10.1068/c0240.

Degen, Mónica Montserrat, and Gillian Rose. *The New Urban Aesthetic: Digital Experiences of Urban Change*. London: Bloomsbury, 2022.

de Hoyos, Arturo. "Masonic Rites and Systems." In *Handbook of Freemasonry*. Edited by Henrik Bogdan and Jan A. M. Snoek, 355–377. Leiden, Netherlands: Brill, 2014.

de Lange, Michiel. "The Smart City You Love to Hate: Exploring the Role of Affect in Hybrid Urbanism." In *Proceedings of The Hybrid City II: Subtle rEvolutions*, edited by Dimitris Charitos, Iouliani Theona, Daphne Dragona, and Charalampos Rizopoulos. Athens, Greece, 2013. Accessed April 11, 2015. https://www.researchgate.net/publication/310476534_The_smart_city_you_love_to_hate_Exploring_the_role_of_affect_in_hybrid_urbanism.

Deleuze, Gilles. *The Fold: Leibniz and the Baroque*. Translated by Tom Conley. Minneapolis: University of Minnesota Press, 1993.

Derrida, Jacques. *Of Grammatology*. Translated by Gayatri Chakravorty Spivak. Baltimore, MD: Johns Hopkins University Press, 1976.

Descartes, Rene. *Discourse on Method and the Meditations*. Translated by F. E. Sutcliffe. Harmondsworth, Middlesex: Penguin, 1968.

Desmedt, Yvo, Shuang Hou, and Jean-Jacques Quisquater. "Audio and Optical Cryptography." In *ASIACRYPT'98, LNCS 1514*, edited by K. Ohta and D. Pei, 392–404. Berlin: Springer-Verlag, 1998.

de Vries, Alex. "Bitcoin's Growing Energy Problem." *Joule* 2 (2018): 801–809.

Dewan, Surbhi, and Latika Singh. "Use of Blockchain in Designing Smart City." *Smart and Sustainable Built Environment and Behaviour* 9, no. 4 (2020): 695–709.

Dey, Swarup, Chunhai Fan, Kurt V. Gothelf, Jiang Li, Chenxiang Lin, Longfei Liu, Na Liu, Minke A. D. Nijenhuis, Barbara Saccà, Friedrich C. Simmel, Hao Yan, and Pengfei Zhan. "DNA Origami." *Nature Reviews* 1, no. 13 (2021): 1–24.

Dickie, John. *The Craft: How the Freemasons Made the Modern World*. London: Hodder and Stoughton, 2020.

Dickie, John. "What the Freemasons Taught the World about the Power of Secrecy." *Time*, August 13, 2020. Accessed February 20, 2021. https://time.com/5877435 /freemason-secrecy/.

Diffie, Whitfield, and Martin Hellman. "New Directions in Cryptography." *IEEE Transactions on Information Theory* IT-22, no. 6 (1976): 644–654.

Dingledine, Roger, Paul Syverson, and Nick Mathewson. "Browse Privately. Explore Freely." Tor. Accessed April 22, 2021. https://www.torproject.org/.

DiResta, Renee, Kris Shaffer, Becky Ruppel, David Sullivan, Robert Matney, Ryan Fox, Jonathan Albright, and Ben Johnson. *The Tactics and Tropes of the Internet Research Agency*. Washington, DC: Senate Select Committee on Intelligence, 2018.

Diver, Laurence. *Digisprudence: Code as Law Rebooted*. Edinburgh: Edinburgh University Press, 2021.

Dobraszczyk, Paul, Carlos López Galviz, and Bradley L. Garrett. *Global Undergrounds: Exploring Cities Within*. London: Reaktion, 2016.

Douglas, Mary. *Purity and Danger: An Analysis of Concepts of Pollution and Taboo*. London: Routledge and Kegan Paul, 1966.

Douglass, Cheryl. "An Introduction to IPFS." *INFURA*, September 17, 2020. Accessed April 28, 2021. https://blog.infura.io/an-introduction-to-ipfs/.

Dounas, Theodoros, Davide Lombardi, and Wassim Jabi. "Framework for Decentralised Architectural Design BIM and Blockchain Integration." *International Journal of*

Architectural Computing 19, no. 2 (2020): 1–17. https://doi.org/10.1177/1478077 120963376.

DuPont, Quinn. "The Printing Press and Cryptography: Alberti and the Dawn of a Notational Epoch." In *A Material History of Medieval and Early Modern Ciphers*, edited by Katherine Ellison and Susan Kim, 95–117. London: Routledge, 2017.

Eco, Umberto. *The Name of the Rose*. Translated by William Weaver. London: Vintage, 1980.

Eco, Umberto. *Reflections on the Name of the Rose*. Translated by William Weaver. London: Secker and Warburg, 1989.

Edwards, John. "Bitcoin's Price History." *Investopedia*, July 2, 2022. Accessed April 10, 2020. https://www.investopedia.com/articles/forex/121815/bitcoins-price-history.asp.

Eliade, Mercea. *The Two and the One*. Translated by J. M. Cohen. London: Harvill Press, 1965.

Ellison, Katherine. "Deciphering and the Exhaustion of Recombination." In *A Material History of Medieval and Early Modern Ciphers*, edited by Katherine Ellison and Susan Kim, 180–207. London: Routledge, 2017.

Ellison, Katherine. *A Material History of Early Modern English Cryptography Manuals*. London: Routledge, 2017.

Elnagar, Randa. "IMF Executive Board Concludes 2021 Article IV Consultation with El Salvador." *International Monetary Fund Press Release No 22/13*, January 25, 2022. Accessed May 14, 2022. https://www.imf.org/en/News/Articles/2022/01/25 /pr2213-el-salvador-imf-executive-board-concludes-2021-article-iv-consultation.

Engin, Z., J. van Dijk, T. Lan, P. A. Longley, P. Treleaven, M. Batty, and A. Penn. "Data-driven Urban Management: Mapping the Landscape." *Journal of Urban Management* 9, no. 2 (2020): 140–150. https://doi.org/10.1016/j.jum.2019.12.001.

Enns, Anthony. "Spiritualist Writing Machines: Telegraphy, Typtology, Typewriting." *Communication +1* 4, no. 1 (2015): 1–27.

Epner, Mia. "Encryption and Public Keys." *Khan Academy*. Accessed March 16, 2021. https://www.khanacademy.org/computing/code-org/computers-and-the-internet /internet-works/v/the-internet-encryption-and-public-keys.

ETA. "Letter Frequencies." Accessed October 24, 2020. http://letterfrequency.org/.

Evans, Robin. "Figures, Doors and Passages." In *Translations from Drawing to Building and Other Essays*, edited by Robin Evans, 55–92. London: Architectural Association, 1997.

Evans, Robin. "Towards Anarchitecture." In *Translations from Drawing to Building and Other Essays*, edited by Robin Evans, 11–34. London: Architectural Association, 1997.

Everett, Hugh. "'Relative State' Formulation of Quantum Mechanics." *Reviews of Modern Physics* 29, no. 3 (1957): 454–462.

Falahat, Somaiyeh. *Re-imaging the City: A New Conceptualisation of the Urban Logic of the "Islamic City."* Berlin: Springer, 2014.

Ferguson, Niels, and Bruce Schneier. *Practical Cryptography*. Indianapolis, IN: Wiley, 2003.

Figueiredo, Sergio M., Sukanya Krishnamurthy, and Torsten Schroeder. *Architecture and the Smart City*. London: Routledge, 2019.

Foucault, Michel. *Discipline and Punish: The Birth of the Prison*. London: Penguin, 1977.

Frankel, Eddy. "An Anonymous Left-wing Art Group Known in the 1990s as Luther Blissett Are Wondering What They Have Unwittingly Helped Create." *The Art Newspaper*, January 19, 2021. Accessed January 30,. www.theartnewspaper.com/2021/01 /19/qanon-the-italian-artists-who-may-have-inspired-americas-most-dangerous -conspiracy-theory.

Fraser, Murray. "Design Research in Architecture, Revisited." In *Artistic Research: Charting a Field in Expansion*, edited by Paulo de Assis and Lucia D'Errico, 128–145. London: Rowman and Littlefield, 2019.

Freud, Sigmund. "The 'Uncanny.'" In *The Penguin Freud Library, Volume 14: Art and Literature*. Edited by Albert Dickson, 335–376. Harmondsworth, UK: Penguin, 1990.

Freudenthal, Hans. *Lincos: Design of a Language for Cosmic Intercourse*. Amsterdam: North Holland, 1960.

Furjan, Helene. *Glorious Visions: John Soane's Spectacular Theatre*. London: Routledge, 2011.

Furno, Angelo, Marco Fiore, Razvan Stanica, Cezary Ziemlicki, and Zbigniew Smoreda. "A Tale of Ten Cities: Characterizing Signatures of Mobile Traffic in Urban Areas." *IEEE Transactions on Mobile Computing: Institute of Electrical and Electronics Engineers* 16, no. 10 (2017): 2682–2696.

Gadamer, Hans-Georg. *Truth and Method*, 2nd rev. ed. Translated by Joel Weinsheimer and Donald G. Marshall. New York: Continuum, 2004. Originally published in German in 1960.

Gaffarel, James. *Unheard-of Curiosities Concerning the Talismanical Sculpture of the Persians; The Horoscope of the Patriarkes; and the Reading of the Stars*. Translated by Edmund Chilmead. ProQuest, 1650.

Galton, Francis. "Intelligible Signals between Neighbouring Stars." *Fortnightly Review* 60, no. 1 November (1896): 657–664.

Geddes, Patrick. *Cities in Evolution: An Introduction to the Town Planning Movement and to the Study of Civics*. London: Williams and Norgate, 1915.

Gehl, Jan. *Cities for People*. Washington, DC: Island Press, 2010.

Gell-Mann, Murray. *The Quark and the Jaguar: Adventures in the Simple and the Complex*. London: Abacus, 1995.

Gibson, Arthur, and Niamh O'Mahony. *Ludwig Wittgenstein: Dictating Philosophy to Francis Skinner—The Wittgenstein-Skinner Manuscripts*. Cham, Switzerland: Springer Nature Switzerland, 2020.

Gibson, James J. *The Ecological Approach to Visual Perception*. Boston: Houghton Mifflin, 1979.

Gibson, James J. *The Senses Considered as Perceptual Systems*. London: Allen and Unwin, 1966.

Giuliani, Rudy W. "Election Theft of the Century | Rudy Giuliani | Ep. 84." *Rudy Giuliani's Common Sense*, November 6, 2020. Accessed November 6, 2020. https://www.youtube.com/watch?v=NVSJriRbxQQ&feature=youtu.be.

Gleick, James. *Chaos: Making a New Science*. London: Heinemann, 1988.

Glock, Hans-Johann. *A Wittgenstein Dictionary*. Oxford: Blackwell, 1996.

Godbout, Jacques T. *The World of the Gift*. Translated by Donald Winkler. Montreal: McGill-Queen's University Press, 1998.

Gold, Jon, and Keith Shaw. "What Is Edge Computing and Why Does It Matter?" *Network World*, June 29, 2021. Accessed August 11, 2021. https://www.networkworld.com/article/3224893/what-is-edge-computing-and-how-it-s-changing-the-network.html.

Golumbia, David. "Bitcoin as Politics: Distributed Right-wing Extremism." In *Moneylab Reader: An Intervention in Digital Economy*, edited by Geert Lovink, Nathaniel Tkacz, and Patricia de Vries, 117–131. Amsterdam: Institute of Network Cultures, 2015.

Gorleé, Dinda L. *Wittgenstein's Secret Diaries: Semiotic Writing in Cryptography*. London: Bloomsbury Academic, 2020.

Graham, Stephen D. N. "Software-Sorted Geographies." *Progress in Human Geography* 29, no. 5 (2005): 562–580.

Graham, Stephen, and Simon Marvin. *Splintering Urbanism: Networked Infrastructures, Technological Mobilities and the Urban Condition*. London: Routledge, 2001.

Gray, Colin M., Yubo Kou, Bryan Battles, Joseph Hoggatt, and Austin L. Toombs. "The Dark (Patterns) Side of UX Design." In *CHI '18: Proceedings of the 2018 CHI*

Conference on Human Factors in Computing Systems, 1–14. Montréal, Canada: ACM, 2018.

Greenberg, Andy. "How Not to Prevent a Cyberwar with Russia." *Wired*, June 18, 2019. Accessed June 26, 2019. https://www.wired.com/story/russia-cyberwar-escalation -power-grid/.

Greenfield, Adam, and Kim Nurri. *The City Is Here for You to Use*. Part 1: *Against the Smart City*. New York: Do Projects, 2013.

Greenwald, Glenn. "Edward Snowden: The Whistleblower Behind the NSA Surveil-lance Revelations." *The Guardian*, June 11, 2013. Accessed November 27, 2019. https:// www.theguardian.com/world/2013/jun/09/edward-snowden-nsa-whistleblower -surveillance.

Greenwald, Glenn. "NSA Collecting Phone Records of Millions of Verizon Custom-ers Daily." *The Guardian*, June 6, 2013. Accessed November 27, 2019. https://www .theguardian.com/world/2013/jun/06/nsa-phone-records-verizon-court-order.

Gregory, Bruce. *Inventing Reality: Physics as Language*. New York: Wiley, 1988.

Gusev, Marjan, and Schahram Dustdar. "Going Back to the Roots—The Evolution of Edge Computing, an IoT Perspective." *IEEE Internet Computing* 22, no. 2 (2018): 5–15. https://doi.org/10.1109/MIC.2018.022021657.

Haines, Avril . *Foreign Threats to the 2020 US Federal Elections*. Washington DC: National Intelligence Council, 2021.

Hall, Edward T. *The Hidden Dimension*. Garden City, NY: Anchor Books, 1966.

Hall, Matt, and John Watkinson. "CryptoPunks." *Larva Labs*. Accessed April 21, 2021. https://www.larvalabs.com/cryptopunks.

Hamilton, Eric. "What Is Edge Computing: The Network Edge Explained." *Cloud-wards*, December 27, 2018. Accessed October 15, 2021. https://www.cloudwards.net /what-is-edge-computing/.

Harari, Yuval Noah. *Sapiens: A Brief History of Humankind*. London: Penguin, 2011.

Hawking, Stephen. *A Brief History of Time: From the Big Bang to Black Holes*. New York: Bantam, 1988.

Heft, Harry. "Affordances and the Perception of Landscape: An Inquiry into Envi-ronmental Perception and Aesthetics." In *Innovative Approaches to Researching Land-scape and Health: Open Space—People Space 2*, edited by Catharine Ward Thompson, Peter Aspinall, and Simon Bell, 9–32. London: Routledge, 2010.

Heisenberg, Werner. *Physics and Philosophy: The Revolution in Modern Science*. London: Penguin, 1958.

Hexeosis. "About Crypto Art: Crypto Art, as a Concept, Is Digital Art That Is Signed and Authenticated by the Artist and Gallery." Accessed January 23, 2021. http://hexeosis.com/cryptoart#:~:text=The%20terms%20of%20SuperRare%20state,Can%20the%20artwork%20be%20resold%3F.

Hexeosis. "Why I Sell My Gifs as Crypto Art." *The Art of Animated GIFs*, May 4, 2020. Accessed January 23, 2021. https://the-art-of-animated-gifs.tumblr.com/post/617190492181053440/hexeosis-why-i-sell-my-gifs-as-crypto-art.

Heyworth, Robin. "Monte Alban: The Encrypted City." *Uncovered History*, April 16, 2014. Accessed September 23, 2017. https://uncoveredhistory.com/mexico/monte-alban-the-encrypted-city/.

Hillier, Bill. *Space Is the Machine: A Configurational Theory of Architecture*. London: Space Syntax, University College London, 2007.

Hnilica, Sonja. "The Metaphor of the City as a Thinking Machine: A Complicated Relationship and Its Backstory." In *Architecture and the Smart City*, edited by Sergio M. Figueiredo, Sukanya Krishnamurthy, and Torsten Schroeder, 68–83. London: Routledge, 2019.

Hodge, Rae. "What Is Tor? A Beginner's Guide to Using the Private Browser." *dearJulius.com*, May 22, 2020. Accessed July 13, 2022. https://tech.dearjulius.com/2020/05/what-is-tor-a-beginners-guide-to-using-the-private-browser.html.

Hodges, Andrew. *Alan Turing: The Enigma of Intelligence*. London: Unwin Paperbacks, 1985.

Hoffman, Hugo. *Cybersecurity Bible: Security Threats, Frameworks, Cryptography & Network Security*. Hugo Hoffman, 2020.

Hofstadter, Douglas R., and Daniel C. Dennett. *Gödel, Escher, Bach: An Eternal Golden Braid*. New York: Basic Books, 1979.

Hollander, Justin B. "Why NFTs Matter to Urban Planning." *Planetizen*, April 14, 2021. Accessed November 7, 2021. https://www.planetizen.com/features/112959-why-nfts-matter-urban-planning.

Homer. *The Odyssey*. Translated by Walter Shewring. Oxford: Oxford University Press, 1980. Written ca. 750 BC.

Howard, Philip N., Bharath Ganesh, Dimitra Liotsiou, John Kelly, and Camille François. *The IRA, Social Media and Political Polarization in the United States, 2012–2018*. Oxford: Computational Propaganda Research Project, University of Oxford, 2018.

Howe, Daniel C., Vincent Toubiana, and Helen Nissenbaum. "TrackMeNot." Accessed April 23, 2021. https://trackmenot.io.

Howe, Daniel C., Mushon Zer-Aviv, and Helen Nissenbaum. "AdNauseam." Accessed April 23, 2021. https://adnauseam.io.

Huber, Martin Fritz. "How to Make GPS Art." *Outside*, June 1, 2015. Accessed July 13, 2022. https://www.outsideonline.com/health/running/how-make-gps-art/

Hughes, Mark. "How Elliptic Curve Cryptography Works." *All About Circuits*, June 26, 2019. Accessed May 15, 2021. https://www.allaboutcircuits.com/technical -articles/elliptic-curve-cryptography-in-embedded-systems/.

Huizinga, Johan. *Homo Ludens: A Study of the Play Element in Culture*. Boston: Beacon Press, 1955.

Hyde, Lewis. *Trickster Makes This World: Mischief, Myth and Art*. New York: North Point Press, 1998.

Ito, Mizuko, Daisuke Okabe, and Misa Matsuda, eds. *Personal, Portable, Pedestrian: Mobile Phones in Japanese Life*. Cambridge, MA: MIT Press, 2006.

Jacobs, Jane. *The Death and Life of Great American Cities*. New York: Random House, 1993.

Jeon, Woojay, Maxwell Jordan, and Mahesh Krishnamoorthy. "On Modeling Asr Word Confidence." *Cornell University*, June 2, 2020. Accessed October 15, 2020. https://arxiv.org/abs/1907.09636v2.

Jones, John Chris. *Design Methods: Seeds of Human Futures*. London: Wiley, 1970.

Joss, Simon, Frans Sengers, Daan Schraven, Federico Caprotti, and Youri Dayot. "The Smart City as Global Discourse: Storylines and Critical Junctures across 27 Cities." *Journal of Urban Technology* 26, no. 1 (2019): 3–34. https://doi.org/10.1080/10630732 .2018.1558387.

Jung, Carl G., M.-L. von Franz, Joseph L. Henderson, Jolande Jacobi, and Aniela Jaffe. *Man and His Symbols*. London: Picador, 1978.

Kahn, David. *The Codebreakers: The Story of Secret Writing*. New York: Scribner, 1996. Kindle.

Kahn, Douglas. *Noise, Water, Meat: A History of Sound in the Arts*. Cambridge, MA: MIT Press, 2001.

Kamczycki, Artur. "The Kabbalistic Alphabet of Libeskind: The Motif of Letter-Shaped Windows in the Design of the Jewish Museum in Berlin." *Ikonotheka* 28 (2008): 7–40.

Kaminska, Izabella. "From Bitcoin to QAnon: Bits to Qbits." *Financial Times: FT Alphaville blog*, August 26, 2021. Accessed January 30, 2021. https://www.ft.com /content/59c03a8e-2b6e-45d2-bf28-744391ffa372.

Karpel, Mark A. "Family Secrets: I. Conceptual and Ethical Issues in the Relational Context II. Ethical and Practical Considerations in Therapeutic Management." *Family Process* 19 (1980): 295–306.

Kessler, Glenn. "Trump's Day on Twitter: Living in an Immaterial World." *Washington Post: Fact Checker*, November 17, 2020. Accessed November 21, 2020. https://www.washingtonpost.com/politics/2020/11/17/trumps-day-twitter-living-an-immaterial-world/.

Kharpal, Arjun. "A Major Chinese Bitcoin Mining Hub Is Shutting Down Its Cryptocurrency Operations." *CNBC*, March 2, 2021. Accessed March 24, 2021. https://www.cnbc.com/2021/03/02/china-bitcoin-mining-hub-to-shut-down-cryptocurrency-projects.html.

Kiayias, Aggelos, A. Russell, B. David, and R. Oliynykov. "Ouroboros: A Provably Secure Proof-of-Stake Blockchain Protocol." In *Annual International Cryptology Conference*, 357–388. Amsterdam: Springer, 2017.

Kim, Krista. "Mars House: The First NFT Digital Home in the World." *SuperRare*, March 15, 2021. Accessed March 24, 2021. https://editorial.superrare.co/2021/03/15/mars-house-nft-digital-home/.

Kirsanov, Vladimir. "Leibniz in Paris." In *The Global and the Local: The History of Science and the Cultural Integration of Europe. Proceedings of the 2nd ICESHS (September 6–9)*, edited by M. Kokowski, 353–364. Cracow, Poland: International Conference of the European Society for the History of Science, 2006.

Kitchin, Rob. "From a Single Line of Code to an Entire City: Reframing the Conceptual Terrain of Code/Space." In *Code and the City*, edited by Rob Kitchin and Sung-Yueh Perng, 15–26. London: Routledge, 2016.

Kittler, Friedrich, Dorothea von Mücke, and Philippe L. Similon. "Gramophone, Film, Typewriter." *October* 41 (1987): 101–118.

Knoop, Douglas, G. P. Jones, and Douglas Hamer. *The Early Masonic Catechisms. QUATUOR CORONATI LODGE*. Accessed February 21, 2021. https://theeducator.ca/wp-content/uploads/2018/02/The-Early-Masonic-Catechisms-by-Harry-Carr.pdf.

Knuth, Donald. *The Art of Computer Programming. Volume 4 Fascicle 0: Introduction to Combinatorial Algorithms and Boolean Functions*. Upper Saddle River, NJ: Addison-Wesley, 2008.

Knuth, Donald E. "Two Thousand Years of Combinatorics." In *Combinatorics: Ancient and Modern*, edited by Robin Wilson and John J. Watkins, 3–37. Oxford: Oxford University Press, 2013.

Kurokawa, Kishō. *Metabolism in Architecture*. London: Studio Vista, 1977.

LaBelle, Brandon. *Background Noise: Perspectives on Sound Art*. New York: Continuum, 2006.

Lake, Josh. "What Is AES Encryption and How Does It Work?" *Comparitech*, February 17, 2020. Accessed March 16, 2021. https://www.comparitech.com/blog /information-security/what-is-aes-encryption/.

Lakoff, George, and Mark Johnson. *Metaphors We Live By*. Chicago: University of Chicago Press, 1980.

Langewiesche, William. "Welcome to the Dark Net, a Wilderness Where Invisible World Wars Are Fought and Hackers Roam Free." *Vanity Fair*, September 11, 2016. Accessed June 11, 2017. http://www.vanityfair.com/news/2016/09/welcome -to-the-dark-net.

Lawrence, Paul. "Graffiti Strategy for Edinburgh." *Culture and Communities Committee*, September 15, 2020. Accessed March 21, 2021. https://democracy.edinburgh .gov.uk/documents/s26152/Graffiti%20Strategy%20for%20Edinburgh-FINAL.pdf.

Lebeuf, Arnold. "The Alphabet and the Sky." In *The Inspiration of Astronomical Phenomena VI*, edited by Enrico Maria Corsini, 317–326. Vol. 441 of ASP Conference Series. San Francisco: Astronomical Society of the Pacific, 2011.

Lefebvre, Henri. *Rhythmanalysis: Space, Time and Everyday Life*. Translated by Stuart Elden and Gerald Moore. London: Continuum, 2004.

Leibniz, Gottfried Wilhelm. "Monadology." In *Philosophical Texts*, 268–281. Oxford: Oxford University Press, 1998.

Lessig, Lawrence. *Code and Other Laws of Cyberspace*. New York: Basic Books, 1999.

Levin, Thomas Y., Ursula Frohne, and Peter Weibel, eds. *CTRL [SPACE]: Rhetorics of Surveillance from Bentham to Big Brother*. Cambridge, MA: MIT Press, 2002.

Levy, Steven. "Crypto Rebels." *Wired*, February 1, 1993. Accessed November 30, 2021. https://www.wired.com/1993/02/crypto-rebels/.

Lewis, Antony. *The Basics of Bitcoins and Blockchains: An Introduction to Cryptocurrencies and the Technology that Powers Them*. Coral Gables, FL: Mango, 2018.

Lewis, Antony. "A Gentle Introduction to Blockchain Technology." *Bits on Blocks: Thoughts on Blockchain Technology*, September 9, 2015. Accessed July 28, 2017. https:// bitsonblocks.net/2015/09/09/a-gentle-introduction-to-blockchain-technology/.

Lewis, Antony. "A Gentle Introduction to Immutability of Blockchains." *Bits on Blocks: Thoughts on Blockchain Technology*, February 29, 2016. Accessed July 28, 2017. https:// bitsonblocks.net/2016/02/29/a-gentle-introduction-to-immutability-of-blockchains/.

Lindon, Stanton J., ed. *The Alchemy Reader: From Hermes Trismegistus to Isaac Newton*. Cambridge: Cambridge University Press, 2003.

Lipman, Andrew D., Alan D. Sugarman, and Robert F. Cushman. *Teleports and the Intelligent City*. Homewood, IL: Dow Jones-Irwin, 1986.

Loo, Becky P. Y., Feiyang Zhang, Janet H. Hsiao, Antoni B. Chan, and Hui Lan. "Applying the Hidden Markov Model to Analyze Urban Mobility Patterns: An Interdisciplinary Approach." *Chinese Geographical Science* 31, no. 1 (2021): 1–13. 10.1007/s11769-021-1173-0

Look, Brandon C. "Gottfried Wilhelm Leibniz." In *Stanford Encyclopedia of Philosophy*, edited by Edward N. Zalta. https://plato.stanford.edu/archives/spr2020/entries/leibniz/2020.

Loos, Adolf. *Ornament and Crime*. London: Penguin, 2019.

Lowenthal, David. *The Past Is a Foreign Country*. Cambridge: Cambridge University Press, 1985.

Lynch, Kevin. *The Image of the City*. Cambridge, MA: Technology Press, 1960.

Lynch, Kevin, and Gary Hack. *Site Planning*. Cambridge, MA: MIT Press, 1984.

Lyons, Glenn. "Getting Smart about Urban Mobility: Aligning the Paradigms of Smart and Sustainable." *Transportation Research Part A* 115 (2016): 4–14. http://dx.doi.org/10.1016/j.tra.2016.12.001.

Lyons, James. "English Letter Frequencies." *Practical Cryptography*. Accessed October 23, 2020. http://practicalcryptography.com/cryptanalysis/letter-frequencies-various-languages/english-letter-frequencies.

Mac Sithigh, Daithi, and Mathias Siems. "The Chinese Social Credit System: A Model for Other Countries?" *The Modern Law Review* 82, no. 6 (2018): 1034–1071.

Madeira, Antonio. "How Does a Hashing Algorithm Work?" *CryptoCompare*, March 13, 2019. Accessed June 13, 2020. https://www.cryptocompare.com/coins/guides/how-does-a-hashing-algorithm-work/.

Maier, Jessica. *Rome Measured and Imagined: Early Modern Maps of the Eternal City*. Oxford: Oxford University Press, 2015.

Manaugh, Geoff. *A Burglar's Guide to the City*. New York: Farrar, Straus and Giroux (Macmillan), 2016.

Manchester, Julia. "Mueller Report Suggests Trump Intended to Obstruct Investigation, Says Ex-Watergate Prosecutor." *The Hill*, April 23, 2019. Accessed August 8, 2019. https://thehill.com/hilltv/rising/440192-evidence-in-mueller-report-suggests-trump-had-intent-to-obstruct-probe-says-ex.

Mann, Kathryn. "The Science of Encryption: Prime Numbers and Mod *n* Arithmetic." Accessed May 3, 2021. https://math.berkeley.edu/~kpmann/encryption.pdf.

Marcus, Joyce. "How Monte Albán Represented Itself." In *The Art of Urbanism: How Mesoamerican Kingdoms Represented Themselves in Architecture and Imagery*, edited by William Leonard Fash and Leonardo López Luján, 77–106. Cambridge, MA: Harvard University Press, 2009.

Maria-Reina Bravo, Doris. "Doris Salcedo, Shibboleth." *Khan Academy*. Accessed April 13, 2020. https://www.khanacademy.org/humanities/ap-art-history/global -contemporary-apah/21st-century-apah/a/doris-salcedo-shibboleth.

Martin, Keith. *Cryptography: The Key to Digital Security, How It Works, and Why It Matters*. New York: Norton, 2020.

Martin, Rod A., and Thomas Ford. *The Psychology of Humor: An Integrative Approach*. Cambridge, MA: Academic Press, 2018.

Marx, Karl. *Karl Marx: Selected Writings*. Oxford: Oxford University Press, 1977.

Mattern, Shannon. *A City Is Not a Computer: Other Urban Intelligences*. Princeton, NJ: Princeton University Press, 2021.

Mattern, Shannon. *Code and Clay, Data and Dirt: Five Thousand Years of Urban Media*. Minneapolis: University of Minnesota Press, 2017.

Matthews, W. H. "Mazes and Labyrinths: A General Account of Their History and Development (1922)." *Project Gutenberg EBook*, July 2014. Accessed February 27, 2021. https://www.gutenberg.org/files/46238/46238-h/46238-h.htm.

May, Tim. "The Crypto Anarchist Manifesto" (1988). Accessed July 13, 2022 http:// nakamotoinstitute.org/crypto-anarchist-manifesto/#selection-51.6-87.64.

McCall Smith, Alexander. *44 Scotland Street*. London: Abacus 2005.

McDermott, Fiona. "New Sensorial Vehicles: Navigating Critical Understandings of Autonomous Futures." In *Architecture and the Smart City*, edited by Sergio M. Figueiredo, Sukanya Krishnamurthy, and Torsten Schroeder, 247–256. London: Routledge, 2019.

McEwen, Indra. *Vitruvius: Writing the Body of Architecture*. Cambridge, MA: MIT Press, 2003.

McLuhan, Marshall. *The Gutenberg Galaxy: The Making of Typographic Man*. Toronto: University of Toronto Press, 1962.

Merkle, Ralph C. "A Digital Signature Based on a Conventional Encryption Function." In *Proceedings of Advances in Cryptology Conference—CRYPTO '87*, vol. 293 (Lecture Notes in Computer Science), edited by Carl Pomerance, 369–378. Berlin: Springer, 1988.

MI5. "Introduction to Bulk Data." Gathering Intelligence. Accessed November 27, 2019. https://www.mi5.gov.uk/bulk-data.

Mitchell, William J. *The Reconfigured Eye: Visual Truth in the Post-Photographic Era*. Cambridge, MA: MIT Press, 2001.

Mitchell, William J. *Placing Words: Symbols, Space, and the City*. Cambridge, MA: MIT Press, 2005.

Moody, Glyn. "What's the Difference Between 'Mass Surveillance' and 'Bulk Collection'? Does It Matter?" *Techdirt*, January 20, 2016. Accessed November 27, 2019. https://www.techdirt.com/articles/20160115/09582933351/whats-difference-between -mass-surveillance-bulk-collection-does-it-matter.shtml.

More, Thomas. *Utopia*. Translated by Paul Turner. Harmondsworth, UK: Penguin, 1965.

Mueller, Robert. *Report on the Investigation into Russian Interference in the 2016 Presidential Election*. Washington, DC: U.S. Department of Justice, 2019.

Mumford, Lewis. *The Culture of Cities*. New York: Open Road Media, 2016.

Naderzadeh, Ali. *Cryptocurrency Mining Using Renewable Energy: An Eco-Innovative Business Model*. Fork Mining FMT Whitepaper, May 28, 2021. Accessed November 5, 2021. https://forkmining.com/FMTFiles/FMT-WhitePaper.pdf.

Nakamoto, Satoshi. "Bitcoin: A Peer-to-Peer Electronic Cash System." *Bitcoin*, n.d. Accessed June 19, 2017. https://bitcoin.org/bitcoin.pdf.

Nance, Malcolm. *The Plot to Betray America: How Team Trump Embraced Our Enemies, Compromised Our Security and How We Can Fix It*. New York: Hachette Books, 2019.

Nance, Malcolm. *They Want to Kill Americans: The Militias, Terrorists, and Deranged Ideology of the Trump Insurgency*. New York: St Martin's Press, 2022.

Naor, M., and A. Shamir. "Visual Cryptography." *Workshop on the Theory and Application of Cryptographic Techniques* 950 (1995): 1–12.

Negarestani, Reza. *Cyclonopedia: Complicity with Anonymous Materials*. Melbourne: re.press, 2008.

Neuman, Fredric. "Keeping Secrets: A Bad Idea That Colors Much of Life." *Psychology Today*, January 30, 2013. Accessed March 19, 2021. https://www.psychologytoday .com/us/blog/fighting-fear/201301/keeping-secrets?collection=123315.

Nevejan, Caroline, Pinar Sefkatli, and Scott Cunningham. *City Rhythm*. Delft: TU, 2018.

Newell, A., and H. A. Simon. *Human Problem Solving*. Englewood Cliffs, NJ: Prentice-Hall, 1972.

Nicholson, Scott. "A Recipe for Meaningful Gamification." In *Gamification in Education and Business*, edited by Torsten Reiners and Lincoln C. Wood, 1–22. New York: Springer, 2014.

Nietzsche, Friedrich Wilhelm. *The Will to Power*. Translated by Walter Kaufmann, and Reginald John Hollongdale. New York: Vintage Books, 1968.

Nietzsche, Friedrich Wilhelm. *The Nietzsche Reader*. Translated by R. J. Hollingdale. London: Penguin, 1977.

Norman, Donald A. *The Design of Everyday Things*. New York: Basic Books, 2002.

Nöth, Winfried. *Handbook of Semiotics*. Bloomington: Indiana University Press, 1990.

Oberhaus, Daniel. *Extraterrestrial Languages*. Cambridge, MA: MIT Press, 2019.

Odgers, Jo, Flora Samuel, and Adam Sharr. *Primitive: Original Matters in Architecture*. Abingdon, Oxon: Routledge, 2006.

Ohlman, Herbert. "Subject-Word Letter Frequencies with Applications to Superimposed Coding." In *Proceedings of the International Conference on Scientific Information: Two Volumes*, 903–916. Washington, DC: National Academies Press, 1959.

Okrand, Marc. *The Klingon Dictionary: The Official Guide to Klingon Words and Phrases*. New York: Pocket Books, 1992.

Oliva, Gustavo A., Ahmed E. Hassan, and Zhen Ming (Jack) Jiang. "An Exploratory Study of Smart Contracts in the Ethereum Blockchain Platform." *Empirical Software Engineering* 25 (2020): 1864–1904. https://doi.org/1810.1007/s10664-10019 -09796-10665.

Oliver, G. *The Pythagorean Triangle: The Science of Numbers*. London: John Hogg, 1875.

Ong, Walter J. *Orality and Literacy: The Technologizing of the Word*. London: Routledge, 2002.

Ong, Walter J. *Ramus: Method, and the Decay of Dialogue from the Art of Discourse to the Art of Reason*. New York: Octagon, 1972.

Ong, Walter J. *Rhetoric, Romance, and Technology: Studies in the Interaction of Expression and Culture*. Ithaca, NY: Cornell University Press, 1971.

Packard, Vance. *The Hidden Persuaders*. Brooklyn, NY: Ig Publishing, 2007. First published in 1957.

Palladio, Andrea. *The Four Books of Architecture*. New York: Dover, 1965. First published in English in 1737.

Pazaitisa, Alex, Primavera De Filippib, and Vasilis Kostakis. "Blockchain and Value Systems in the Sharing Economy: The Illustrative Case of Backfeed." *Technological Forecasting & Social Change* 125 (2017): 105–115.

Penn, Alan, and Kinda Al Sayed. "Spatial Information Models as the Backbone of Smart Infrastructure." *Environment and Planning B: Urban Analytics and City Science* 44, no. 2 (2017): 197–203. https://doi-org.ezproxy.is.ed.ac.uk/10.1177/2399808317693478.

Pennick, Nigel. *Mazes and Labyrinths*. London: Robert Hale, 1990.

Perlroth, Nicole. *This Is How They Tell Me the World Ends: The Cyberweapons Arms Race*. New York: Bloomsbury, 2021.

Perlroth, Nicole, and Scott Shane. "In Baltimore and Beyond, a Stolen N.S.A. Tool Wreaks Havoc." *New York Times*, May 25, 2019. Accessed June 5, 2019. https://www.nytimes.com/2019/05/25/us/nsa-hacking-tool-baltimore.html.

Permissionjunkie. "What Secret Codes Are All around Us That Only People 'in the Know' Recognize?" *Reddit*. Accessed March 10, 2021. https://www.reddit.com/r/AskReddit/comments/3x004o/what_secret_codes_are_all_around_us_that_only/.

Petrauskiene, Vilma, and Loreta Saunoriene. "Application of Dynamic Visual Cryptography for Optical Control of Chaotic Oscillations." *Journal of Vibroengineering Procedia* 15 (2017): 81–87.

Plato. *The Republic of Plato*. Translated by Francis MacDonald Cornford. London: Oxford University Press, 1941.

Plotinus. *The Essence of Plotinus: Extracts from the Six Enneads and Porphyry's Life of Plotinus*. Translated by Stephen Mackenna. New York: Oxford University Press, 1948.

Poovey, Mary. *A History of the Modern Fact: Problems of Knowledge in the Science of Wealth and Society*. Chicago: University of Chicago Press, 1998.

Popov, Vesselin, Michal Kosinski, David Stillwell, and Bartosz Kielczewski. "Apply Magic Sauce: Trait Prediction Engine." *University of Cambridge Psychometrics Centre*. Accessed June 28, 2018. https://applymagicsauce.com/.

Prescott, Andrew. "The Old Charges." In *Handbook of Freemasonry*, edited by Henrik Bogdan and Jan A. M. Snoek, 33–49. Leiden, Netherlands: Brill, 2014.

Priani, Ernesto. "Raymon Llull." *Stanford Encyclopedia of Philosophy* (Winter 2016 Edition). Accessed October 30, 2021. https://plato.stanford.edu/entries/llull/.

Prominski, Martin, and Spyridon Koutroufinis. "Folded Landscapes: Deleuze's Concept of the Fold and Its Potential for Contemporary Landscape Architecture." *Landscape Journal* 28, no. 2 (2009): 151–165.

Pullan, Wendy. "Agon in Urban Conflict: Some Possibilities." In *Phenomenologies of the City: Studies in the History and Philosophy of Architecture*, edited by Henriette Steiner and Maximilian Sternberg, 213–224. Farnham, UK: Ashgate, 2015.

Pullman, Philip. *Northern Lights: His Dark Materials*. London: Scholastic, 2017. First published in 1995.

Rabiner, L. R., and B. H. Juang. "An Introduction to Hidden Markov Models." *IEEE Acoustics, Speech and Signal Processing Magazine* 3 (1986): 4–16.

Raymond, Eric S. *The Cathedral and the Bazaar: Musings on Linux and Open Source by an Accidental Revolutionary.* Beijing: O'Reilly, 2001.

Reddy, Michael. "The Conduit Metaphor: A Case of Frame Conflict in Our Language about Language." In *Metaphor and Thought,* edited by Andrew Ortony, 284–324. Cambridge: Cambridge University Press, 1979.

Reiff, Nathan. "How to Identify Cryptocurrency and ICO Scams." *Investopedia,* March 28, 2022. Accessed July 13, 2022. https://www.investopedia.com/tech/how-identify-cryptocurrency-and-ico-scams/.

Relph, Edward. "Spirit of Place and Sense of Place in Virtual Realities." *Techné: Research in Philosophy and Technology, Special Issue: Real and Virtual Places* 10, no. 3. (2007): 17–25. http://scholar.lib.vt.edu/ejournals/SPT/v10n3.

Rescher, Nicholas. "Leibniz's *Machina Deciphratoria*: A Seventeenth-Century Proto-Enigma." *Cryptologia* 38, no. 2 (2014): 103–115. https://doi.org/10.1080/01611194.2014.885789.

Rice, Ronald E., Sandra K. Evans, Katy E. Pearce, Anu Sivunen, Jessica Vitak, and Jeffrey W. Treem. "Organizational Media Affordances: Operationalization and Associations with Media Use." *Journal of Communication* 67, no. 1 (2017): 106–130. https://doi.org/10.1111/jcom.12273.

Rivera, J. W. "Potential Negative Effects of a Cashless Society: Turning Citizens into Criminals and Other Economic Dangers." *Journal of Money Laundering Control* 22, no. 2 (2019): 350–358.

Robinson, Michael D., Ryan L. Boyd, and Adam K. Fetterman. "An Emotional Signature of Political Ideology: Evidence from Two Linguistic Content-coding Studies." *Personality and Individual Differences* 71 (2014): 98–102.

Roma N. "Animated Stereogram." YouTube. Accessed October 5, 2021. https://www.youtube.com/watch?v=IZpsbQMQFBs.

Ross, George M. "Leibniz and the Nuremberg Alchemical Society." *Studia Leibnitiana* 6, no. 2 (1974): 222–248.

Rovelli, Carlo. "Relational Quantum Mechanics." *International Journal of Theoretical Physics* 35, no. 8 (1996): 1637–1678.

Rowe, Colin, and Fred Koetter. *Collage City.* Cambridge, MA: MIT Press, 1978.

Rumelhart, D. E., and J. L. McClelland, eds. *Parallel Distributed Processing: Explorations in the Microstructure of Cognition.* Cambridge, MA: MIT Press, 1987.

Saleh, Fahad. "Blockchain without Waste: Proof-of-Stake." *The Review of Financial Studies* 34 (2021): 1156–1190.

Schade, Christian D. *Free Will and Consciousness in the Multiverse: Physics, Philosophy, and Quantum Decision Making*. Cham, Switzerland: Springer Nature Switzerland AG, 2018.

Schiff, Adam. *Midnight in Washington: How We Almost Lost Our Democracy and Still Could*. New York: Penguin Random House, 2021.

Schneier, Bruce. *Beyond Fear: Thinking Sensibly About Security in an Uncertain World*. New York: Copernicus Books, 2003.

Scholem, Gershom G. *Major Trends in Jewish Mysticism*. London: Thames and Hudson, 1955.

Sebeok, Thomas A. *Signs: An Introduction to Semiotics*. Toronto: University of Toronto Press, 1999.

Seetharam, Anand. "RSA (Rivest, Shamir, Adleman) Algorithm Explained With Example." *CSEdu4All*, January 29, 2021. Accessed May 4, 2021. https://www.youtube.com/watch?v=KPkm2yvyGi8.

Sennett, Richard. *The Craftsman*. London: Penguin, 2008. Kindle.

Shamir, Adi. "How to Share a Secret." *Communications of the ACM* 22, no. 11 (1979): 612–613.

Shannon, Claude E., and William Weaver. *The Mathematical Theory of Communication*. Urbana: The University of Illinois Press, 1963.

Sharr, Adam. *Heidegger's Hut*. Cambridge, MA: MIT Press, 2006.

Sheldon, Mark D. "Auditing the Blockchain Oracle Problem." *Journal of Information Systems* 35, no. 1 (2021): 121–133. DOI:10.2308/ISYS-19-049

Shields, Jennifer. *Collage and Architecture*. New York: Routledge, 2013.

Shiller, Robert J. "The Bitcoin Narratives." In *Narrative Economics: How Stories Go Viral and Drive Major Economic Events*, 3–11. Princeton, NJ: Princeton University Press, 2019.

Shojaei, Alireza, Ian Flood, Hashem Izadi Moud, Mohsen Hatami, and Xun Zhang. "An Implementation of Smart Contracts by Integrating BIM and Blockchain." In *FTC 2019: Proceedings of the Future Technologies Conference—Advances in Intelligent Systems and Computing*, vol 1070, edited by K. Arai, R. Bhatia, and S. Kapoor, 519–527. Cham, Switzerland: Springer, 2019.

Sigalos, MacKenzie. "El Salvador Has Just Started Mining Bitcoin Using the Energy from Volcanoes." *Crypto Decoded* (CNBC), October 2, 2021. Accessed July 13, 2022. https://www.cnbc.com/2021/10/01/el-salvador-just-started-mining-bitcoin-with-volcanoes-for-the-first-time-ever-and-theyve-already-made-269.html.

Simon, Herbert. *The Sciences of the Artificial*. Cambridge, MA: MIT Press, 1969.

Simon, Herbert. "The Structure of Ill-Structured Problems." *Artificial Intelligence* 4 (1973): 181–201.

Simpson, John, and Edmund Weiner, eds. *The Oxford English Dictionary*. 2nd ed. Oxford, UK: Oxford University Press, 2022.

Singh, Simon. *The Code Book: The Secret History of Codes and Code-Breaking*. London: Fourth Estate, 2000.

Smith, Robert. "Quantum Instruction Set—Computerphile." *Computerphile*. Accessed March 27, 2021. https://www.youtube.com/watch?v=ZN0lhYU1f5Q.

Snodgrass, Adrian, and Richard Coyne. *Interpretation in Architecture: Design as a Way of Thinking*. London: Routledge, 2006.

Snoek, Jan A. M. "Masonic Rituals of Initiation." In *Handbook of Freemasonry*, edited by Henrik Bogdan and Jan A. M. Snoek, 321–327. Leiden, Netherlands: Brill, 2014.

Snowden, Edward. *Permanent Record*. London: Macmillan, 2019.

Sommerville, Ian. *Software Engineering*. Harlow, UK: Addison-Wesley, 2001.

Steadman, Philip. *Architectural Morphology: An Introduction to the Geometry of Building Plans*. London: Pion, 1983.

Stephenson, Neal. *Snow Crash*. New York: Bantam Books, 1992.

Stevenson, David. "The Origins of Freemasonry: Scotland." In *Handbook of Freemasonry*, edited by Henrik Bogdan and Jan A. M. Snoek, 50–62. Leiden, Netherlands: Brill, 2014.

Stiny, George. "Kindergarten Grammars: Designing with Froebel's Building Gifts." *Environment and Planning B: Planning and Design* 7 (1980): 409–462.

Stone, Oliver. *Snowden*. Open Road Films, 2016.

Suiche, Matt. "Shadow Brokers: NSA Exploits of the Week." *Comae Technologies*, August 15, 2016. Accessed July 25, 2022. https://medium.com/comae/shadow -brokers-nsa-exploits-of-the-week-3f7e17bdc216.

Sullivan, Nick. "A (Relatively Easy to Understand) Primer on Elliptic Curve Cryptography." *The Cloudflare Blog*, October 23, 2013. Accessed May 6, 2021. https://blog .cloudflare.com/a-relatively-easy-to-understand-primer-on-elliptic-curve-cryptography/.

Sundararajan, Arun. *The Sharing Economy: The End of Employment and the Rise of Crowd-Based Capitalism*. Cambridge, MA: MIT Press, 2016.

Surowiec, Paweł. "Post-truth Soft Power: Changing Facets of Propaganda, *Kompromat*, and Democracy." *Georgetown Journal of International Affairs* 18, no. 3 (2017): 21–27. https://doi.org/10.1353/gia.2017.0033.

Sutton, Sam. *Cryptocurrency Mining: The Ultimate Guide to Understanding Bitcoin, Ethereum, and Litecoin Mining*. Scotts Valley, CA: Createspace Independent Publishing Platform, 2018.

Szabo, Nick. "Formalizing and Securing Relationships on Public Networks." *First Monday* 2, no. 9 (1997). https://firstmonday.org/ojs/index.php/fm/article/view/548.

Tanizaki, Junichiro. *In Praise of Shadows*. Translated by Thomas J. Harper and Edward G. Seidensticker. Stony Creek, CT: Leete's Island Books, 1977. First published in Japanese in 1933.

Tanna, Sunil. *Codes, Ciphers, Steganography and Secret Messages*. High Wycombe, UK: Answers 2000 Ltd, 2020.

Tavernise, Sabrina, and David Yaffe-Bellany. "Death of a Crypto Company." *The Daily: The New York Times*. Podcast, July 25, 2022. Accessed July 26, 2022. https://www.nytimes.com/2022/07/25/podcasts/the-daily/cryptocurrency-celsius-network.html?.

Taylor, Luke. "The World's First Bitcoin Republic." *New Scientist* 253, no. 3376 (2022): 14. https://doi.org/10.1016/S0262-4079(22)00360-8.

Thiel, Philip. "Processional Architecture." *Ekistics* 17, no. 103 (1964): 410–413.

Topo, Päivi, Helinä Kotilainen, and Ulla Eloniemi-Sulkava. "Affordances of the Care Environment for People with Dementia—An Assessment Study." *Health Environments Research and Design Journal* 5, no. 4 (2012): 118–138.

Topolinski, Sascha, and Rolf Reber. "Gaining Insight into the 'Aha' Experience." *Current Directions in Psychological Science* 19, no. 6 (2010): 402–405.

Townsend, Robert M. *Distributed Ledgers: Design and Regulation of Financial Infrastructure and Payment Systems*. Cambridge, MA: MIT Press, 2020.

Tschumi, Bernard. *Architecture and Disjunction*. Cambridge, MA: MIT Press, 1994.

Turner, Alasdair, and Alan Penn. "Encoding Natural Movement as an Agent-Based System: An Investigation into Human Pedestrian Behaviour in the Built Environment." *Environment and Planning B: Planning and Design* 29 (2002): 473–490.

Turner, Victor. *The Forest of Symbols: Aspects of Ndembu Ritual*. Ithaca, NY: Cornell University Press, 1967.

Underwood, Dudley. *Numerology, or, What Pythagoras Wrought*. Cambridge: Cambridge University Press, 1998.

Unger, Stefanie, and Walter Grassl. "Insta-Holidays and Instagrammability." *Journal of Tourism, Leisure and Hospitality* 2, no. 2 (2020): 92–103.

Vaidhyanathan, Siva. *The Googlization of Everything: (And Why We Should Worry)*. Berkeley: University of California Press, 2011.

van Ditmar, Delfina Fantini. "The IdIoT in the Smart Home." In *Architecture and the Smart City*, edited by Sergio M. Figueiredo, Sukanya Krishnamurthy, and Torsten Schroeder, 157–164. London: Routledge, 2019.

Vidler, Anthony. *The Writing of the Walls: Architectural Theory in the Late Enlightenment*. Princeton, NJ: Princeton Architectural Press, 1987.

Vilenkin, Alexander, and Max Tegmark. "The Case for Parallel Universes: Why the Multiverse, Crazy as It Sounds, Is a Solid Scientific Idea." *Scientific American*, July 19, 2011. Accessed November 20, 2020. https://www.scientificamerican.com/article/multiverse-the-case-for-parallel-universe/.

Vitruvius, Pollio. *Vitruvius: Ten Books on Architecture*. Translated by Morris Hicky Morgan. New York: Dover Publications, 1960. Written ca. 50 AD.

Watkin, David. "Freemasonry and Sir John Soane." *Journal of the Society of Architectural Historians* 54, no. 4 (1995): 402–417.

Watson, Jon. "Famous Codes and Ciphers through History and Their Role in Modern Encryption." *Comparitech*, May 13, 2017. Accessed March 16, 2021. https://www.comparitech.com/blog/information-security/famous-codes-and-ciphers-through-history-and-their-role-in-modern-encryption/.

Weinberg, Justin. "Philosophers on GPT-3 (updated with replies by GPT-3)." *Daily Nous*, July 30, 2020. Accessed June 2, 2022. https://dailynous.com/2020/07/30/philosophers-gpt-3.

Weinsheimer, Joel C. *Gadamer's Hermeneutics: A Reading of Truth and Method*. New Haven and London: Yale University Press, 1985.

Weston, Dagmar. "The Lantern and the Glass: On the Themes of Renewal and Dwelling in Le Corbusier's Early Art and Architecture." In *Spirituality and the City*, edited by Iain Boyd Whyte, 146–177. London: Routledge, 2003.

Whitson, Jennifer R. "Gaming the Quantified Self." *Surveillance and Society* 11, no. 1/2 (2013): 163–176.

Wicker, Stephen B. "The Ethics of Zero-Day Exploits—The NSA Meets the Trolley Car." *Communications of the ACM* 64, no. 1 (2021): 97–103.

Wilkins, Emily J., Derek Van Berkel, Hongchao Zhang, Monica A. Dorning, Scott M. Beck, and Jordan W. Smith. "Promises and Pitfalls of Using Computer Vision to Make Inferences about Landscape Preferences: Evidence from an Urban-proximate Park System." *Landscape and Urban Planning* 219 (2022): 104315. https://doi.org/10.1016/j.landurbplan.2021.104315.

Wilkins, John. *Mercury: Or the Secret and Swift Messenger*. London: I. Norton for John Maynard and Timothy Wilkins, 1641.

Williams, Kim, Lionel March, and Stephen R. Wassell. *The Mathematical Works of Leon Battista Alberti*. Basel: Springer Basel, 2010.

Williams, Sarah. *Data Action: Using Data for Public Good*. Cambridge, MA: MIT Press, 2020.

Willis, Katharine S., and Alessandro Aurigi. *Digital and Smart Cities*. Abingdon, UK: Routledge, 2018.

Windsor, Luke. "An Ecological Approach to Semiotics." *Journal for the Theory of Social Behaviour* 34, no. 2 (2004): 179–198.

Wittgenstein, Ludwig. *Philosophical Investigations*. Translated by G. E. M. Anscombe. Oxford: Blackwell, 1953.

Wittgenstein, Ludwig. *Tractatus Logico Philosophicus*. Translated by C. K. Ogden. London: Routledge and Kegan Paul, 1922.

Wright, Mark, Henrik Ekeus, Richard Coyne, James Stewart, Penny Travou, and Robin Williams. "Augmented Duality: Overlapping a Metaverse with the Real World." In *Proceedings of the International Conference on Advances in Computer Entertainment Technology, ACE 2008, December 3–5*, edited by Masa Inakage and Adrian David Cheok, 263–266. Yokahama, Japan: ACM, 2008.

Yoo, Wonsang, Yu Eun, and Jaemin Jung. "Drone Delivery: Factors Affecting the Public's Attitude and Intention to Adopt." *Telematics and Informatics* 35, no. 6 (2018): 1687–1700.

Young Rival. "Black Is Good [Official] (Autostereogram Video)." YouTube, January 27, 2014. Accessed October 5, 2021. https://www.youtube.com/watch?v=2AKtp3XHn38.

Zahid, Muniza, Byeonghoon Kim, Rafaqat Hussain, Rashid Amin, and Sung Ha Park. "DNA Nanotechnology: A Future Perspective." *Nanoscale Research Letters* 8, no. 119 (2013): 1–13. https://doi.org/10.1186/1556-276X-8-119.

Zhang, Yinan, Fei Wang, Chao Jie, Mo Xie, Huajie Liu, Muchen Pan, Enzo Kopperger, Xiaoguo Liu, Qian Li, Jiye Shi, Lihua Wang, Jun Hu, Lianhui Wang, Friedrich C. Simmel, and Chunhai Fan. "DNA Origami Cryptography for Secure Communication." *Nature Communications* 10, no. 5469 (2019): 1–8. https://doi.org/10.1038/s41467-019-13517-3.

Zuboff, Shoshana. *The Age of Surveillance Capitalism: The Fight for a Human Future at the New Frontier of Power*. London: Profile Books, 2019.

Index